Legal Basis of Global Tissue Banking

A Proactive Clinical Perspective

Legal Basis of Global Tissue Banking

A Proactive Clinical Perspective

Editor

Glyn O Phillips

Research Transfer Ltd, UK

Phillips Hydrocolloids Research Ltd, UK

World Scientific

NEW JERSEY • LONDON • SINGAPORE • BEIJING • SHANGHAI • HONG KONG • TAIPEI • CHENNAI

Published by

World Scientific Publishing Co. Pte. Ltd.
5 Toh Tuck Link, Singapore 596224
USA office: 27 Warren Street, Suite 401-402, Hackensack, NJ 07601
UK office: 57 Shelton Street, Covent Garden, London WC2H 9HE

Library of Congress Cataloging-in-Publication Data
Legal basis of global tissue banking : a proactive clinical perspective / edited by Glyn O. Phillips.
p. ; cm.
Includes bibliographical references and index.
ISBN 978-9814663434 (hardcover : alk. paper) -- ISBN 978-9814663441 (electronic)
I. Phillips, Glyn O., editor.
[DNLM: 1. Tissue Banks--legislation & jurisprudence. 2. Internationality. QS 523]
RD127
362.17'83--dc23
2014048337

British Library Cataloguing-in-Publication Data
A catalogue record for this book is available from the British Library.

Typeset by Stallion Press
Email: enquiries@stallionpress.com

Printed in Singapore

CONTENTS

ABOUT THE EDITOR

Glyn O. Phillips is currently Chairman of Research Transfer Ltd, Phillips Hydocolloids Research Ltd, the Cellucon Trust and the International Hydrocolloids Conferences. He is Visiting Professor at Glyndwr University and a consultant to several industrial organisations. In 2010, the Glyn O. Phillips Hydrocolloids Research Centre was opened in Hubei University of Technology, Wuhan, China, of which Professor Phillips is Chairman.

Professor Phillips graduated with a degree in chemistry from the University of Wales and holds the degrees of Doctor of Philosophy (PhD) and Doctor of Science (DSc) from the University of Wales. He is a chartered chemist and Fellow of the Royal Society of Chemistry. His own University of Wales conferred upon him (in 1999) the degree of Doctor of Laws Honoris Causa (LLD) for "outstanding contributions to science and education". He has been awarded many honours, including the Hopkins Medal of New York for "contributions to science in North America", the Science and Technology Medal (2004) from the National Eisteddfod of Wales "for eminent contributions to scientific literature in Welsh" and has been elected an Honorary Member for Life of the Society of Fiber Science and Technology, Japan, the Latin American Association of Tissue Banking, the Asia Pacific

Association of Surgical Tissue Banking and the European Association of Tissue Banking. In 2012, Professor Phillips was elected Fellow of the Learned Society of Wales.

Professor Phillips' research interests have centred upon carbohydrates and biomaterial systems, particularly on food hydrocolloids and medically related biopolymers of connective tissue. Professor Phillips is now Executive Editor of the journal *Food Hydrocolloids*, and has published 64 books and (currently) 646 papers in leading scientific journals. These include *New Fibers* (a first edition and an updated revised edition, Woodhead Publishing); *Innovation & Technology Transfer in Japan and Europe* (Routledge); the series *Gums and Stabilisers* (17 books); the *Cellucon* series (10 books); *Handbook of Food Proteins* (2011); *Handbook of Hydrocolloids* (second and enlarged edition, 2009); *Hyaluronan*: *Vol* 1 *Chemical, Biochemical and Biological Aspects* and *Vol 2 Biomedical, Medical and Clinical Aspects* (2002, Woodhead Publishing). He is also currently Editor-in-Chief of *Advances in Tissue Banking* and Editor of *International Journal of Cell and Tissue Banking*.

ABOUT THE AUTHORS

Li Baoxing is the current Director of Shanxi Provincial Tissue Bank and Director of Biomaterials Research and Development Centre of the China Institute for Radiation Protection. Because his work in the research and development of irradiated bone allografts obtained First Class in the National Defense Science and Technology Award, he became State Council Expert for Special Allowance. He is also a lifetime member of the American Association of Tissue Banks and the Asia Pacific Association of Surgical Tissue Banking.

Professor Li graduated from Suzhou Medical College as a radiology professional in 1978. Since then he has been engaged in radiation medicine and tissue transplantation materials research. His efforts in making bone allografts obtained State Food and Drug Administration (SFDA) registration, beginning a new era of bone allograft clinic applications in China.

In order to propel forward China's tissue bank programmes, Professor Li presided over the drafting of two industry standards for bone allografts. He is also one of the founders of China Association of Bone Tissue Bank and is Vice Chairman of the organisation. He has twice hosted the National Symposium of Allobone Grafting Technique.

He has written over 100 scientific papers on tissue transplantation and tissue banking and a book entitled *Tissue Allografts Transplantation and Tissue Bank*.

Mario Cabraja is a neurosurgeon specialising in spine surgery, and is the head of an interdisciplinary department for spine diseases consisting of orthopaedic surgeons, neurologists, radiologists and neurosurgeons at the Vivantes Auguste-Viktoria-Hospital in Berlin, Germany.

He underwent his neurosurgical residency at the Charité Medical School in Berlin, received his board certification as a neurosurgeon in 2008 and was a spine fellow of the European Association of Neurosurgical Societies. He absolved his spine fellowship under the tutorial of Professor W.C. Peul at the neurosurgical department of the Leiden University, the Netherlands. In 2009, he became a neurosurgical consultant of the Charité Medical School and has headed the section for spinal neurosurgery of the Charité since 2011. He was awarded a postdoctoral academic degree (Habilitation) for his scientific research on spinal diseases at the Charité Medical School in 2013. At the end of 2013 he left the Charité Medical School to take up his current position at the Vivantes Hospital Network in Berlin.

Daniel Harshman is a trial lawyer at FisherBroyles, LLP in Los Angeles, California. He has over 30 years of experience as a business trial lawyer and has tried cases in Pennsylvania, California, Arizona and Texas. While he specialises in trademark, copyright and trade secret litigation, over his 30-year career, Mr Harshman has handled a wide range of litigation including biological implant litigation, business torts, products liability, business contract disputes and insurance litigation.

Mr Harshman is admitted to practice law in California and Pennsylvania (currently inactive). Mr Harshman earned his Juris Doctor degree from Duquesne University School of Law in Pittsburgh, Pennsylvania. During law school, Mr Harshman was a member of the *Duquesne Law Review* where he served as Recent Decisions Editor. He received the American Jurisprudence Award for Pennsylvania and Federal Practice. He graduated *cum laude* with a Bachelor of Arts degree in Business Administration from the California University of Pennsylvania.

Caroline A. Hartill is Executive Vice President and Chief Scientific Officer for RTI Surgical. She joined RTI in November 2001 and has served as Vice President of Quality Assurance and Regulatory Affairs since 2003. In her role at RTI, Ms Hartill oversees all research and development, quality assurance, regulatory affairs and clinical studies for the company which operates in the United States, Europe and Asia. She has served as a board member for BioFlorida and is Chair for the Tissue Products Workgroup of the Advanced Medical Technology Association (AdvaMed). She is a member of the AATB Board of Governors, has also served for several years as a member and then Chair of the Accreditation Committee of the AATB and is an appointed member of the Agency for Healthcare Administration (AHCA), State of Florida Organ and Tissue Transplant Advisory Committee. Ms Hartill earned a Bachelor's degree with honours in health sciences from Birmingham University Medical School in England, as well as a Master's degree in management from the University of Wolverhampton in England. She has also earned Master's level credits in sterilization science from Manchester University.

Peter Hergert is a lawyer at Paul-Ehrlich-Institut, Federal Institute for Vaccines and Biomedicines, Management Support Section, Langen, Germany.

He received his law degree in 1998 from Johann Wolfgang Goethe University, Frankfurt/Main, Germany. He completed his practical education in 2001 during which he worked in the fields of intellectual property law, criminal law and environmental law, as well as common private and administrative law. He worked as a freelancing associate in a law firm specialised in environmental law until 2002.

Peter Hergert joined Paul-Ehrlich-Institut in 2002 and soon became Deputy Head of the section for legal affairs. In this role he was responsible for legal consultancy of the Institute's management and sections regarding all aspects of the Institute's work, especially pharmaceutical law. He specialises in legal issues regarding blood and blood products, tissues and cells, advanced medicinal products, medical devices and many more subgenres of pharmaceutical law. In 2013, he switched from the lawyers team to the management team and since then he has been involved in more political issues of the Institute. He is now Deputy Head of the management support section.

Deborah Hutchison is an attorney and public health expert with both domestic and international experience. She is currently the Portfolio Director for West and Central Africa at Family Health International (FHI 360), located in Washington DC.

Ms Hutchison obtained her law degree in 2005 from the Catholic University of Leuven in Belgium. She went on to obtain a Masters in Law with a focus on health law and policy from Vanderbilt University Law School in Nashville, Tennessee in 2006. In 2011, Ms Hutchison earned a Masters in Public Health from the University of the Philippines in Manila.

Ms Hutchison has several years of experience as a regulatory consultant for some of the largest private hospital systems and healthcare organisations in the United States. She also has provided strategic planning and regulatory support to US government agencies. Ms Hutchison developed a successful compliance resource centre with a policy library and sanction screening service for private and public sector organisations. Alongside her work in the US healthcare regulatory field, Ms Hutchison has developed expertise in the policy and practice of international public health, serving as an academic with a distinguished public health centre and managing multi-million dollar programmes overseas. In her present role at FHI 360, Ms Hutchison provides oversight to several country offices and ensures compliance with government regulations in implementing public health, education and livelihood programmes overseas.

Eulogia Kairiyama is Manager of Radiation Applications and Technology in the Nuclear Technology Applications Area of the National Atomic Energy Commission of Argentina.

She received her degree in biochemistry from the Buenos Aires University (1968) and her postgraduate degree on bacterial taxonomy and radiation-resistant bacteria studies at the Science University of Tokyo and Takasaki Radiation Chemistry Institute (1978–1979 and 1992, respectively). From 1993 to 2012, she was related to radiation sterilisation and tissue banking activity in national, regional and interregional projects of IAEA as coordinator and counterpart of Argentina. She was Director of IAEA's five Regional Training Courses for Tissue Banking Operation and Radiation Sterilisation of Tissue Allografts (Buenos Aires, from 1999 to 2012) and of the interregional and regional training courses on the IAEA Code of Practice for Radiation Sterilisation of Human Tissues (Buenos Aires, 2004 and 2012).

Jack Klecan has been lead trial counsel in over 225 jury trials in 40 years of courtroom experience. He has been recognised in the publication *Best Lawyers* in both New Mexico and Arizona. He has defended a wide variety of cases in medical malpractice, products liability, professional liability and insurance coverage. In addition, he was lead counsel in a class action involving funding for special education and defended the State of New Mexico in a class action regarding foster care. New Mexico has a tort claims cap for governmental entities. Jack represented the City of Albuquerque on a constitutional challenge to the cap and prevailed before the State Supreme Court.

Mr Klecan collaborated with his co-authors in a medical malpractice trial involving 50 plaintiffs who claimed that the use of polymethyl methacrylate (PMMA) as an intervertebral device was improper. Professor Glyn Phillips explained at trial that PMMA is the most commonly used non-metallic implant and that it has been proven safe in numerous applications. Dr Cabraja explained at trial the experience in Germany using PMMA as an intervertebral device which was proven as safe and effective as other devices.

Helena E. Lazares is an associate in the Philadelphia office of Lewis Brisbois. Helena's primary area of practice is in the life sciences.

During law school, she served as a law clerk at a mid-sized suburban Philadelphia law firm, concentrating on complex civil litigation. She has also served as a certified legal intern for the Villanova Civil Justice Clinic, representing indigent clients in a variety of civil suits.

She received her Juris Doctor (JD) from Villanova University School of Law in 2014. She was elected as Editor of Student Works for the *Villanova Environmental Law Journal* and was a student representative for the Honor Board.

Peter Myint was trained as a veterinary surgeon and later went into research and obtained his PhD in radiation chemistry of connective tissues. Dr Myint has 20 years of experience in tissue banking. As early as in 1990s he developed the operational activities of North Wales and Oswestry Tissue Bank in Wales which became one of the largest tissue banks in the United Kingdom providing bone, tendon, cartilage, meniscus and skin allografts for human application. He served on the British Association for Tissue Banking at the executive level. He also served on the government's policy collaborative group as well as in a working party to develop an implementation plan for regulatory and licensing methods for tissue and cell establishments in the United Kingdom. Dr Myint functioned as a technical expert on the radiation sterilisation of tissues to the International Atomic Energy Agency (IAEA) and continues to be involved with IAEA in collaborative research programmes. Dr Myint is a co-founder of Veterinary Tissue Bank, Europe's first veterinary tissue bank providing allograft tissues and stem cells for orthopaedic procedures in the treatment of dogs and cats.

Aziz Nather is a consultant orthopaedic surgeon in National University Hospital, Singapore and Medical Director of National University Hospital Tissue Bank.

He received his surgical degree in 1978 from University of Edinburgh and University of Glasgow.

He has done clinical work and research on bone allograft transplantation and has written several books published by World Scientific including

Bone Graft and Bone Substitutes (2005),
Radiation in Tissue Banking (2007) and *Procurement, Processing and Transplantation of Bone Allografts* (2010).)

He is currently Director of IAEA/NUS Internet Diploma Training Course for Tissue Bank Operators — a one-year distance learning programme run since 1997.

Jorge Morales Pedraza currently works as an independent consultant on international affairs and possesses degrees in mathematics and economic sciences. Formerly, he was a Cuban ambassador for disarmament and non-proliferation for more than 25 years. He was Deputy and Head of the Cuban delegation to the IAEA General Conference, IAEA Board of Governors and the Plenary Sessions of the Organisation for the Prohibition of Chemical Weapons (OPCW) Preparatory Commission, among other international responsibilities. In the 1980s, Morales was appointed Ambassador and Permanent Representative of Cuba to the IAEA and in the 1990s gained the same title with the OPCW. In addition, he was invited university professor in Mathematics Sciences and an Invited Professor for International Relations in the Diplomatic Academy of Cuba. Throughout the 1990s and into the 2000s Morales Pedraza worked for the IAEA as Regional Project Coordinator for the Latin American region, Interim Section Head for Latin America and Senior Manager in the Director's office. During that period he was in charge of the IAEA program on radiation and tissue banking among other regional and interregional projects. Over the past years, he was involved in the preparation, as author and coauthor, of more than 65 articles published by international publisher's houses, as well as ten chapters for various books. During this period he also authored eight books and was invited editor for international journals. Morales Pedraza is a member of the editorial teams of six specialized international journals.

John T. Salvucci is Chair of Life Sciences at Lewis Brisbois. He regularly associates in the defence of high exposure cases for excess insurers. In this capacity, he is often brought in as a consultant tasked with developing an understanding of the key medical/scientific issues, identifying and retaining preeminent authorities, and formulating the defense strategy. In addition, he serves as national counsel in mass tort cases. In the past he has coordinated the defence of hundreds of claims on behalf of a weight loss company, a medical device manufacturer of bone screws, and a processor of human bone allografts. John and his team currently serve as national coordinating counsel both for a dental management company with affiliated clinics located throughout the United States and for hospitals and clinics in matters related to fungal meningitis arising from steroid injections.

Ralf Reinhard Tönjes is a research scientist and group leader at Paul-Ehrlich-Institut (PEI), Federal Institute for Vaccines and Biomedicines, Division of Medical Biotechnology, Langen, Germany.

He received his MS degree in biology and germanistics (1983) and his PhD degree in biology (1986) from Philipps University, Marburg, Germany. His thesis involved molecular analyses of histone genes and proteins for chromatin studies. As a scientist he worked at Fraunhofer-Institute, Hannover, Germany, and at Rockefeller University, New York (1986–1991). His research interests included the liver-specific regulation of genes in mutant and transgenic mice.

Dr Tönjes has been a section head at PEI since 1991. He is in charge of evaluations of medicinal products including human tissue preparations and xenogeneic cell therapeutics. He founded the German working group for xenotransplantation (DAX) together with Dr Joachim Denner in 1998. He teaches as an adjunct professor for biochemistry at Goethe University, Frankfurt/Main, Germany. His area of research is focused on endogenous retroviruses, retroelements, zoonotic agents and infections in the course of xenotransplantation, as well as human and porcine transcriptomics.

Zhao Yaping is currently Vice-General Manager of Shanxi OsteoRad Biomaterial Co., Ltd. He is the newsletter editor of *Chinese Journal of Bone and Joints.* He is an expert on the Shanxi Expert Committee on Food and Drug Safety.

Professor Zhao graduated in radiation medicine from the Suzhou Medical College, and practised plastic surgery for 15 years at Shanxi Provincial Plastic Surgery Hospital. In 2002, he transferred to Shanxi Provincial Tissue Bank and began research in allogeneic skin graft procedures and storage.

Professor Zhao set up the procedures and standards for alloskin grafts and alloskin products, organised clinical trials of alloskin grafts, and submitted the application of registration of alloskin to the China Food and Drug Administration (CFDA). He drafted two industry standards of allogenic bone grafts in 2006. He contributed to a monograph in *Tissue Allograft Transplantation and Tissue Bank* (2010) and a chapter for the book *Tissue and Cell Donation: An Essential Guide* (2009).

Norimah Yusof is a radiation biologist by training and has always been promoting nuclear power for the benefits of mankind. She spent 38 years as a senior scientist while heading Agrotechnology and Biological Sciences at the Malaysian Nuclear Agency (1973–2011). Although busy with science and technology management in the Agency, she was active in R&D activities at national, regional and international levels, especially in the IAEA-coordinated projects in the fields of food irradiation, radiation sterilisation of medical products, biomaterial and tissue banking. She has produced four patents in biomaterials (hydrogel wound dressing and scaffold), and is a co-author for two books on tissue banking and an author for chapters in books on tissue banking and radiation sterilisation. She has publications in national and international journals and has been appointed as an advisor/consultant at national level and on several IAEA expert missions.

Her involvement in tissue banking started in 1990 when she assisted Universiti Sains Malaysia (USM) to establish a tissue bank and ensure all tissues produced by the local banks were gamma sterilised. She trained scientists and IAEA fellows in her favourite subjects of radiation sterilisation of tissues, i.e., bioburden, D_{10} values, dose mapping, dose selection, and validation and radiation effects. She was an invited lecturer on radiation sterilisation for USM undergraduate courses in tissue banking for several years. She has also been involved in IAEA/NUS Diploma Programme in Tissue Banking in Singapore since 1998 as a lecturer and examiner. Currently, as Adjunct Professor at the University of Malaya, she has improved the quality management of the UMMC Bone Bank and is encouraging advanced research in biomaterial and tissues for postgraduate students.

PREFACE

Ironically, the epoch-breaking developments in immunology led by Professor Peter Medawar during the latter part of the 20th century proved a hindrance rather than a help to the emerging field of tissue banking. Medical doctors and surgeons were, on balance, of the opinion that tissue taken from one person would inevitably be rejected by a different human recipient. The experience with skin, in particular, seemed to further support this view, with grafts used for the treatment of burns, while first appearing to have grafted successfully, were all eventually rejected.

The use of allograft bone at that time was a true "cottage industry", with orthopaedic surgeons keeping pieces of bone recovered during surgery in their personal refrigerators and then using them for skeletal reconstruction.

Towards the end of the 20th century, from 1991 in particular, tissue banking and the clinical utilisation of such tissues grew gradually into a science. Bone, removed of its immunological components, could be grafted successfully and eventually be incorporated due to the in-growth of new bone. Also, from that time tissue banking centres emerged in the United Kingdom, Europe, Latin America and Southeast Asia, and as the century approached its end these grew in experience, size and confidence.

Unfortunately, legislation to support such a growing tissue transplant system using a range of human tissues lagged behind. It was well into the 20th century before the rumblings of the highly bureaucratic European system were heard and attempts were started to coordinate national legal initiatives. Meanwhile, in the United States, the technology grew rapidly into a highly organised industrial system. The Food and Drug Administration struggled to keep up. Other countries followed behind and even today, 25 years since the renaissance of tissue banking, the legal control of tissue banking can best be described as "patchy" throughout the world.

This book, for the first time, documents the current position and is written by practising experts from the various geographical regions. Our hope is that it might point the way to global harmonisation of the principles and practice of the subject and so rationalise them also in its legal management.

Personally, I must thank the International Atomic Energy Agency who allowed me the opportunity to assist in the development of this technology and in setting up tissue banks in at least 30 countries throughout the world. Now ionising radiation is the method of choice for the sterilisation of a wide range of human tissues. During this 20-year period I made friends and found many supporting colleagues. I dedicate this book to them for their patience and friendship.

Glyn O. Phillips

1

INTRODUCTION: FROM A COTTAGE INDUSTRY TO A GLOBAL BUSINESS

Glyn O. Phillips

Although the clinical practice of using bone grafts to repair, replace or supplement the bone stock has a long history, dating back to McEwen in 1881, the production and practices have grown into a truly industrial-scale international business only over the past 10–15 years or so. Human safety considerations led to the articulation of regulatory systems which increasingly became the target of legal challenges as real or apparent shortcomings in the implementation of the practices became more frequent. This book sets out the way in which the current legal approaches to the production and clinical utilisation of human tissue grafts have developed in various parts of the world.

The development of tissue banking throughout the world has been quite uneven and has taken place over a relatively short time scale. The foundation of current tissue banking was set in place, essentially, by a relatively small number of tissue banks. The early pioneers set up organised tissue banks only in the second half of the 20th century. The pace has eventually been set by the United States where production systems and graft quantities have grown exponentially to industrial proportions. Smaller tissue banks in Europe are increasingly being absorbed by their larger counterparts such that the overall production philosophy is spreading internationally. However, the pace at which legal awareness is spreading is not even throughout the world, with the result that certain regions and countries, notably China, have developed their own legal systems. This book describes the situation as currently practised, and will hopefully lead to wider discussions about this subject.

THE DEVELOPMENT OF TISSUE BANKING

When a group of surgeons, in which Geoffrey Burwell from the United Kingdom was a leading figure, showed that frozen preserved allograft could be superior in performance to fresh allogeneic bone, the road was pointed to the more extensive use of bone allografts. However, generally the practice remained a "cottage industry" well into the latter part of the 20th century. This involved orthopaedic surgeons keeping pieces of bone which had been rescued after surgery in individual hospital cold stores. These were usually femoral heads after hip replacement, and these they used as required on an individual basis. There were many exceptions and these surgeons were usually associated with the pioneering tissue banks, which emerged in the 1950s. But progress in the second half of the 20th century was painfully slow. Tissue banking did not easily fit into the medical psyche. While medical education included a consideration of the historical development of transplantation, the message given to the modern era was in many ways counterproductive. Even as the end of the 20th century approached, orthopedic surgeons still had not embraced the new clinical opportunities provided by sterile non-immunogenic allograft bone. There was a legitimate concern about possible disease transfer from contaminated grafts, but there was additionally a deep lack of awareness of the potential of the developing technology. I found this to be so in 1990 even in one of the most famous orthopaedic hospitals in the United Kingdom. A young orthopaedic Registrar who joined our embryonic research tissue bank in North Wales to pursue a higher degree was regarded as an oddity in pursuing such a career path.

Why such ignorance about the potential of these grafts, particularly since many countries such as Poland, for example, from 1975 onwards, had produced thousands of radiation-sterilised allografts which were transplanted successfully every year? Possibly, the medical attitude was unconsciously derived from the experience with skin grafting, which had first suggested that allografts and even xenografts might be grafted permanently and be incorporated into the human body. This we now acknowledge was a degree of self-deception. Any skin graft leaves a pad of collagen when it rejects, and this may have been mistaken for a successful graft. Epithelialisation can also occur under a rejected graft by in-growth from the sides, and this new skin can be confused with successful grafting. These spurious successes must have received encouragement from the undoubted success of using allograft bone, first by William MacEwan (1881) and then by the successful grafting of cornea in 1905 by Edward Zirm. The confusing allograft era eventually came to an end. Scientists like Schone in 1912 and Lexer in 1914 clearly stated that rejection of allografts was inevitable. The Danish scientist Jensen even suggested that this was an immunological event. World War II had a profound effect on science and accelerated progress in the understanding of kidney disease and tissue transplantation. Problems like massive burns in the 1939–1945 war, and new forms of renal failure, produced new clinical problems. Existing therapies were ineffective. In Britain, the Medical Research Council Burns Unit at Glasgow was established. There, Peter Medawar became interested in the problems of skin grafting after seeing a seriously burnt patient in an Oxford hospital. In Glasgow, Medawar and Gibson showed that a second set of skin grafts was rejected more quickly than the first set. They reported this work in 1943. For Gibson, a plastic surgeon, it was the final proof that allografts were of no immediate clinical use. For Medawar, it was evidence that allograft rejection was a major, unexplained immunological phenomenon. Back in Oxford, Medawar repeated the experiment using rabbits and termed the rejection mechanism *actively acquired immunity*. Medawar and colleagues studied this immunological phenomenon intensively and in 1960 Medawar and Burnet shared the Nobel Prize for Medicine and Physiology. We now know that the judgement of Gibson was premature. Human kidney allografts have now been successfully transplanted throughout the world. Indeed, renal transplantation is now the treatment of choice for end-stage renal failure and cornea grafting is so widely practised and successful that it is often forgotten that it is a tissue allograft, albeit a rather special one. Now more than 1,750,000 tissue allografts are transplanted annually in the

United States, and the procedure is a fundamental part of modern health care. Yet the developments up to the early 1990s were painfully slow.

The period when the European Association of Tissue Banks (EATB) started (in 1993) and other national associations such as the British Association of Tissue Banks (BATB) started marks the period of acceleration into the modern tissue banking era. The growth, thereafter, has been exponential in the United States, linear in Europe, linear also in Australasia but with a lower slope, and asymptotical in the Asia and Latin America regions. China was slower in embracing this technology but now is moving at an ever increasing pace as noted in this book.

THE EARLY PIONEERS

The foundation of current tissue banking was set in place, essentially, by a relatively small number of tissue banks. Without entering into the debate about which actually was the first, we should recall the contributions of the Bethesda Naval Tissue Bank in the United States which George Hyatt founded in 1950, with Ken Sell becoming Director in 1965. The impetus was the return of the Korean War casualties who required reconstructive surgery. From this beginning came great contributions in the subject from Michael Strong and Gary Friedlander. The early publication *Tissue Banking for Transplantation*, edited by Sell and Friedlander (1976) and published by Grune and Stratton, New York, was truly a landmark publication.

At about the same time, Rudolph Klen set up the tissue bank in Hradec Kralove in the old Czechoslovakia in 1952. It was a great occasion on 6 July 2006 when a meeting was held to celebrate the contribution of Professor Klen and congratulate him on his 90th birthday. It was a privilege for me to work with Professor Klen to contribute to and translate his previous publication in the Czech language into English. In 1982, another landmark publication by Klen, *Biological Principles of Tissue Banking,* was published. It is a miracle that Klen emerged to all this great work after being imprisoned as a Jew in a concentration camp during the war years. Letters now exist which were sent by him and his sister Hana and their friend Jura (Jiri) Fantl in 1944 from the ghetto to their families in the *Protektorat*. We cannot now imagine their suffering and their resilience.

Rudolph Klen continued to show this resilience throughout his life, for he had to withstand the communist regime which denied him all recognition, which only came in the form of a professorship from his old university at the end of his life.

The Leeds Tissue Bank was set up in 1956 by another great enthusiast, Frank Dexter. Its early history has been described by his student John Kearney (2006). It all started in the Department of Microbiology of the University of Leeds in 1953. Essentially a technician in microbiology at the University of Leeds Medical School, Dexter had a dream of harvesting tissues for transplantation, and together with Professor Geoffrey Burwell proceeded to use tissues by the hundreds for transplantation. It is a tragedy that this early experience was not recorded in detail. In 1965 when a new purpose-built Regional Burns Unit was opened at Pinderfields, Dexter was invited to be its first Head. His enthusiasm and campaigning for more than 20 years led to the establishment of the Yorkshire Regional Tissue Bank, again at Pinderfield Hospital in Wakefield — an unlikely place some thought to have the flagship tissue bank in the United Kingdom. Frank Dexter, its first Head, retired in 1985, and Dr John Kearney became Director of the Yorkshire Regional Tissue Bank in its new building. I am particularly grateful to Frank Dexter for completely unselfishly helping us to set up our North Wales and Oswestry Tissue Bank in the years up to 1990. Both tissue banks were taken over by the UK Blood Service in the early 2000s.

The Charite Hospital Tissue Bank in Berlin in the old East Germany was set up in 1956, and although well-known in the Eastern block, it was hardly known in the West at the time. The great opening up came in 1990 when an East meets West symposium was held in Berlin, organised by Dr Ruediger von Verson, who had joined the Charite Bank in 1981. The fruits very quickly spilled over and it was a resolution of the meeting to set up an association of tissue banks. After a planning meeting held in Marseilles (with Professor Poitou as host) the European Association of Tissue Banks (EATB) came into being in 1993, and Professor von Verson was its President for the first ten years. In Europe reference should be made also to the pioneers of tissue banking in Poland, namely Professors K. Ostrowski, Anna Dziedic-Goclawska and J. Komander. The tissue bank in Warsaw celebrated its 40th anniversary in 2004.

The sterling work of Dr Nicholas Triantafyllou who started and sustained tissue banking in Greece should also be recognised. He was the first, along with the Polish group, to study in a scientific manner the effects of ionizing radiations on bone and to use the method routinely for end-sterilization. He produced bone grafts and as an orthopaedic surgeon also used them.

The role of the International Atomic Energy Agency (IAEA) in promoting tissue banking (Phillips 1998, 1999, 2001, 2002, 2004; Phillips and Morales 2003; Nather *et al.*, 2003) should also be acknowledged. The IAEA has gained more experience and success than any other international organization in establishing tissue banks in developing countries and applying ionizing radiation for sterilizing tissue grafts, which were then used in transplant surgery (in orthopaedic reconstruction, treatment for cancer, trauma and high velocity impact damage), the treatment of burns, leprosy and intractable skin wounds and pressure sore ulcers. Now due to the diligence of Jorge Morales (a contributor to this volume), who was a powerful manager of these programmes for IAEA, we have a complete record of the achievements of the programme and how each of the countries has contributed.

The main achievements can be summarised as follows:

a) Establishing a program of 25 tissue banks in the Asia Pacific and Latin American regions. Now there are more than 112 tissue banks operating in these two regions.
b) Radiation sterilisation of tissue grafts is the established end-sterilisation method in the countries in the participating regions and has led to a reverse transfer technology effect to the United States and Europe.
c) All countries operate the approved quality assurance programmes.
d) The preparation and training in the use of the following operational documents:

- The IAEA Code of Practice for the Radiation Sterilisation of Tissues Allografts: Requirements for Validation and Routine Control.
- IAEA International Standards for Tissue Banking.
- IAEA Public Awareness Strategies for Tissue Banks.

e) The dissemination of knowledge regarding best practices for tissue banking professionals through training courses, fellowships scientific visits and expert missions.
f) The support and sponsor of several international scientific meetings in the participant regions.
g) The establishment of the Asia-Pacific Association of Surgical Tissue Banking (APASTB) and Latin American Association of Tissue Banks (ALABAT).
h) The establishment of two regional training centres, one in Singapore and another in Buenos Aires, and one interregional centre in Singapore. In these centres 261 students have been trained and 166 have graduated.
i) With the financial and technical support of the IAEA, preparation and utilisation of a comprehensive distance learning curriculum for tissue bank operators and medical doctors in English, Korean and Spanish.
j) The transformation of comprehensive distance learning programmes for tissue bank operators and medical doctors into an Internet training course. The realisation of 334 expert missions, the commissioning of 34 national consultants, the support of 161 participants in several meetings organised by the programme, the approval of 44 fellowships and 34 scientific visits.

LEGISLATION

Legislation and standards were non-existent and the newly formed associations had no brief, no competence and no resources to provide training or supervise the progress of the subject. To provide an outlet to this emerging science specifically related to tissue banking, the publication *Advances in Tissue Banking* was started in 1997, followed by the *Journal of Cell and Tissue Banking* in 2002.

This book describes how the subject has moved on from this developing period which has been outlined. The European Union and the United States have set up their own control systems and other regions and countries are setting systems in place appropriate to their needs as described in this book. The subject is still emerging.

REFERENCES

Alvarez I, Morales Petraza J, Saldías MC, Pérez Campos H, Wodowóz O, Acosta M, Vicentino W, Silva W, Rodríguez G, Machín D, Alvarez O (2009) The impact of the International Atomic Energy Agency (IAEA) program on radiation and tissue banking in Uruguay: Development of tissues quality control and quality management system in the National Multi-tissue Bank of Uruguay, *Cell Tissue Bank* **10**, 173–181.

Dexter F (1965) The preservation for surgical transplantation and subsequent formation of a tissue bank, *J Sci Technol* **11**, 149–176.

Dexter F (1967) Freeze dried heart valves and tissue banking, *Biomed Eng* **2**, 308–316.

Dexter F (1976) Tissue banking in England, *Transplant Proc*, 843–848.

Eulogia K, Morales Pedraza J (2009) The impact of the International Atomic Energy Agency (IAEA) program on radiation and tissue banking in Argentina, *Cell Tissue Bank* **10**, 139–142.

Hilmy N, Manjas M, Abbas FB, Morales Pedraza J (2009) The impact of the International Atomic Energy Agency (IAEA) program on radiation and tissue banking in Indonesia, *Cell Tissue Bank* **10**, 103–107.

Kang YK, Yimm CJ, Morales Pedraza J (2009) The impact of the International Atomic Energy Agency (IAEA) program on radiation and tissue banking in Korea, *Cell Tissue Bank* **10**, 109–113.

Kearney JN (2006) Yorkshire regional tissue bank: Circa 59 years of tissue banking, *Cell Tissue Bank* **7**, 259–264.

Klen R (1982) *Biological Principles of Tissue Banking*, Pergamon Press, Oxford (English Editor and Contributor: Phillips GO).

Lobo Gajiwala A, Morales Pedraza J (2009) The impact of the International Atomic Energy Agency (IAEA) program on radiation and tissue banking in India, *Cell Tissue Bank* **10**, 93–101.

Martınez-Pardo ME, Morales Pedraza J, Sa0nchez Ramı0rez O (2009) The impact of the International Atomic Energy Agency (IAEA) program on radiation and tissue banking in Mexico, *Cell Tissue Bank* **10**, 157–165.

Morales Pedraza J (2006) The International Atomic Energy Agency (IAEA) program in radiation and tissue banking: A successful program for developing countries, *Cell Tissue Bank.* **7**, 237–258.

Morales Pedraza J (2006b) The International Atomic Energy Agency (IAEA) program on radiation and tissue banking: A successful program for developing countries. *Cell Tissue Bank* **7**, 237–258.

Morales Pedraza J (2006c) The International Atomic Energy Agency (IAEA) program on radiation and tissue banking: A successful program for developing countries, *Cell Tissue Bank* **7**, 237–258.

Morales Pedraza J (2008) The importance of the selection of the audiences and the organization of media events within public awareness strategies for tissue banks, *Cell Tissue Bank* **9**, 317–322.

Morales Pedraza J (2011a) A model of a code of ethics for tissue banks operating in developing countries. *Cell Tissue Bank*, doi: 10.1007/s10561-0(2011)11-9279-8.

Morales Pedraza J (2011b) The use of the Internet training course modality in the field of tissue banking: The International Atomic Energy Agency (IAEA)/ National University of Singapore (NUS) experience, *Cell Tissue Bank* **12**, 163–170.

Morales Pedraza J, Martınez-Pardo ME (2012) A review of the International Atomic Energy Agency (IAEA) international standards for tissue banks, *Cell Tissue Bank* **13**, 15–25.

Morales Pedraza J, Phillips GO (2009) The impact of the International Atomic Energy Agency (IAEA) program on radiation and tissue banking in Asia and the Pacific and the Latin American regions, *Cell Tissue Bank* **10**, 79–86.

Morales Pedraza J, Sánchez Noda EO, Lorenzo R, Cardona R, Otero I (2009) The impact of the International Atomic Energy Agency (IAEA) program on radiation and tissue banking in Cuba. *Cell Tissue Bank* **10**, 149–152.

Morales Pedraza J, Vajaradul Y, Alvarez I (2011) The future role of the professional associations in the promotion of tissue banking activities in Asia and the Pacific and in the Latin America regions, *Cell Tissue Bank* **12**, 319–327.

Nather A (2000) Diploma training for technologists in tissue banking, *Cell Tissue Bank* **1**, 41–44.

Nather A, Phillips GO, Morales Pedraza J (2003a) IAEA/NUS distance learning diploma training course for tissue bank operators: Past present and future, *Cell Tissue Bank*, Special Issue, **4**(2–4), 2.

Nather A, Philips GO, Morales Pedraza J (2003b) IAEA/NUS distance learning diploma training course for tissue bank operators: Past, present and future, *Cell Tissue Bank* **4**, 77–84.

Nather A, Phillips GO, Morales Pedraza J, Lee CCW (2009) Training tissue bank operators: The International Atomic Energy Agency (IAEA)/National University of Singapore (NUS) 10 years of experience, *Cell Tissue Bank* **10**, 183–190.

Phillips GO (1999) Tissue banking in the Asia Pacific region, in Phillips GO (ed.) *Advances in Tissue Banking*, Vol. 3, World Scientific, Singapore, pp. 399–402

Phillips GO (2000a) The future role of the International Atomic Energy Authority (IAEA) in tissue banking, *Cell Tissue Bank* **1**, 27–40.

Phillips GO (2000c) Hudson Silva: The pioneer of eye banking in Asia and the Pacific region, *Cell Tissue Bank* **1**, 171–172.

Phillips GO (2000d) The future role of the International Atomic Energy Authority (IAEA) in tissue banking, *Cell Tissue Bank* **1**, 27–40.

Phillips GO (2003a) *Allografts in Bone Healing: Biology and Clinical Applications*, Vol. 1, *Bone Biology and Healing*, World Scientific, Singapore.

Phillips GO (2003b) *Allografts in Bone Healing: Biology and Clinical Applications*, Vol. 2, *Morphogenic Protein and Collagen*, World Scientific, Singapore.

Phillips GO (2005d) *Allografts in Bone Healing: Biology and Clinical Applications*, Vol. 3, *Clinical Applications of Bone Allografts and Substitutes*, World Scientific, Singapore.

Phillips GO, Morales J (2002) Catalysts for better health care, Int Atom Energy Agency Bull **44**(1), 17–20.

Phillips GO, Morales J (2003) The International Atomic Energy Agency (IAEA) programme in radiation and tissue banking: Past, present and future, *Cell Tissue Bank*, Special Issue **4**(2–4), 67–76.

Phillips GO, Tallentire A, Triantafylou N (eds.) (1978) Radiation sterilization of irradiated tissues and their potential uses, The North East Wales Institute, Wales.

Phillips GO, Strong D (1997) The contribution of the International Atomic Energy Agency to tissue banking, *Am Assoc Tissue Banks Tissue Cell Report* **4**(1), 5–10.

Phillips GO, Tatsuzaki H (1998) The tissue banking programme supported by the International Atomic Energy Agency (IAEA), *Transfusion Today*, **37** (December), 24–25.

Phillips GO (ed.) (2006) *Cell Tissue Bank*, Vol. 7(4): special issue to coincide with the merger with *Advances in Tissue Banking*.

Phillips GO (ed. in chief) (1997) *Advances in Tissue Banking*, Vol. 1. World Scientific, Singapore.

Phillips GO (ed. in chief) (1998) *Advances in Tissue Banking*, Vol. 2, World Scientific, Singapore.

Phillips GO (ed. in chief) (1999) *Advances in Tissue Banking*, Vol 3, World Scientific, Singapore.

Phillips GO (ed. in chief) (2000) *Advances in Tissue Banking*, Vol 4, World Scientific, Singapore.

Phillips GO (ed. in chief) (2001) *Advances in Tissue Banking*, Vol. 5, World Scientific, Singapore.

Phillips GO (ed. in chief) (2002) *Advances in Tissue Banking*, Vol. 6, World Scientific, Singapore.

Phillips GO (ed. in chief) (2004) *Advances in Tissue Banking*, Vol. 7, World Scientific, Singapore. Papers relating to the IAEA Tissue Banking Programme published in *J Cell Tissue Bank*.

Vajaradul Y, Morales Pedraza J (2009) The impact of the International Atomic Energy Agency (IAEA) program on radiation and tissue banking in Thailand, *Cell Tissue Bank* **10**, 125–132.

Yusof, N, Morales Pedraza, J (2009) The impact of the International Atomic Energy Agency (IAEA) program on radiation and tissue banking in Malaysia, *Cell Tissue Bank* **10**, 115–117.

2

LEGAL FRAMEWORK FOR INTERNATIONAL OPERATION OF TISSUE BANKS

Peter Myint

INTRODUCTION

Over the last decade there has been a steady increase in tissue banking activities on a global scale. In line with this, regulatory requirements have been introduced by the relevant authorities, and inspection and compliances are now an integral part of the tissue banking practice. This area of tissue banking will play an increasingly important role in international tissue banking as the movement of tissues across borders increases.

As the legal aspects of tissue banking take on a more important role, the development of standards and regulatory requirements continue to improve. The level of regulatory needs and compliances varies from location to location and often the voluntary standards and statutory requirements operate side by side. The regulations tend to centre on the areas that are considered "higher risk" in terms of disease transmission such as donor evaluation, testing and adverse event reporting, and include the guidelines and code of practices which focus on the technical and scientific areas. This chapter summarises the various regulatory requirements that currently operate in various regions and highlights the areas that are significant.

EUROPE

In Europe the European Union Tissue and Cells Directives (EUTCD) sets out to establish a harmonised approach to the regulation of tissues and cells across Europe and set a benchmark for the standards that must be met when carrying out any activity involving tissues and cells for human application. All member states are required to designate a competent authority to implement the requirements of the EUTCD and to ensure that tissue establishments are accredited, authorised or licensed.

The EUTCD is made up of three directives. The Parent Directive (2004/23/EC) provides the framework legislation and sets the standard of quality and safety for the donation, procurement, testing, processing, preservation, storage and distribution of human tissues and cells (Directive 2004/23). The directive does not prevent a member state from introducing or maintaining more stringent protective measures.

The first Technical Directive (2006/17/EC) provides the detailed requirements for donation, procurement and testing of cells and tissues (Directive 2006/17/EC). Subsequently a commission directive (2012/39/EU) was published to amend some technical requirements about the HTLV testing (Directive 2012/39/EU).

The second Technical Directive (2006/86/EC) provides the detailed requirements for traceability, notification of serious adverse reaction, coding, processing, preservation, storage and distribution of human tissues and cell (Directive 2006/86/EC).

The competent authorities are required to provide the European Commission an annual report about the notification of serious adverse reactions and events. The competent authorities are also required to send annual activity data to the European Registry for Organs, Cells and Tissues (EUROCET) which publishes the data on their website www.eurocet.org.

UNITED KINGDOM

The first move towards a formal approach to inspection of tissue banks was in 2001 when the Department of Health published a document "A code of practice for tissue banks" (Department of Health, February 2001).

It formed the basis of the voluntary accreditation scheme applied to tissue banks in the public sector supplying human tissues for therapeutic purposes to the health service. Tissue banks were inspected by the Medicines Control Agency (MCA) which later became Medicines and Healthcare Products Regulatory Agency (MHRA).

At the same time the EUTCD was being developed by the European Commission and the UK Department of Health formed a policy collaborative group to support the development of the directives. This also coincided with the development of the new legislations in the UK regarding the removal, storage, use and disposal of human bodies, organs and tissue. This arose as a consequence of events in 1999–2000 where organs of children were removed, stored and used without proper consent (The Royal Liverpool Children's Inquiry Report; The Report of the Public Inquiry into Children's Heart Surgery at the Bristol Royal Infirmary 1984–1995). It was established that such practices had been widespread from both adults and children without proper consent. A consultation document was launched in July 2002 and after much public consultation, the Human Tissue Act 2004 was passed (Human Tissue Act 2004). The Human Tissue Act 2004 replaced the Human Tissue Act 1961, the Anatomy Act 1984 and the Human Organ Transplant Act 1989. Consent is the fundamental principal of the legislation and underpins the lawful removal, storage and use of body parts, organs and tissue. There is a separate legislation in Scotland — the Human Tissue (Scotland) Act 2006.

The Human Tissue Act 2004 established the Human Tissue Authority (HTA) to regulate the activities concerning the removal, storage, use and disposal of human tissue in 2005. The HTA is a UK competent authority to regulate a wide range of sectors including anatomy, research, transplants, DNA, public display, human application, organ donation and transplant, removal licence and post-mortem. All establishments that are involved in any of the above activities require a licence from the HTA, and there are more than 800 establishments that are licensed by the HTA (see www.hta.gov.uk).

Tissue banking falls under the Human Application sector of the HTA's remit. The HTA licenses and inspects tissue establishments that undertake the procurement, testing, processing, storage, distribution, import and export of tissues and cells for human application under the Human Tissue (Quality and Safety for Human Application) Regulations 2007 (Q&S Regulations) (Human Tissue 2007 No. 1523). The Q&S Regulations 2007 implements the EUTCD and the implementation was fully completed into UK law on 5 July 2007.

Under these Q&S Regulations the licensed establishments are required to meet the standards which are detailed in the HTA's "Guide to Quality and Safety Assurance for Human Tissues and Cells for Patient Treatment" (Human Tissue Authority (2010) Guide to Quality and Safety Assurance for Human Tissues and Cells for Patient Treatment). This is implemented by Directions 003/2010 (Directions given under the Human Tissue Act 2004) which consolidate and clarify the standards. A working guide to compliance with the regulations is provided in the "Quality and Safety Standards" (Human Application: HTA Standards) which is also used as a measure against the standards when tissue establishments are inspected. In addition to the standards there are also a set of Codes of Practices. The relevant Codes to tissue banking include consent (HTA Code of Practice 1, Consent), disposal of human tissue (HTA Code of Practice 5, Disposal of Human Tissue) and import and export of human body, body parts and tissue (HTA Code of Practice 8, Import and Export of Human Bodies, Body Parts and Tissue).

At the heart of the licensing need is storage. Any tissue establishment that stores tissue or cells for human application for more than 48 hours needs a licence to store. Other activities such as testing, procurement, processing, import–export and distribution can take place under a licence or under a third-party agreement with a licensed tissue establishment. Within licences Designated Individual (DI) is a key player under whose supervision the licensed activity is authorised to be carried out and who carries the legal responsibility. The licence application form is evaluated to assess the performance against four sets of standards namely consent; governance and quality system; premises, facilities and equipment; and disposal. The inspection process is a combination of desk-based assessment and site visit inspection. In addition to the licence application, the HTA may ask to submit Preparation Process Dossier (PPD) (HTA Preparation Process Dossier) to validate the processes which are used to prepare tissue and cells to ensure they are fit for the purpose.

Under the Q&A Regulations, an import licence is required to import tissue or cells from outside the European Economic Area (EEA) states or within the EEA if the supplier is not a licensed establishment by the respective competent authority. Likewise, an export licence is needed to supply tissue or cells to outside the EEA or to supply to an unlicensed establishment within the EEA. Under the licensing standard, all licensed tissue establishments are required to submit their annual activity data to the HTA including the import and export figures.

Reporting Requirements

Tissue establishments are required to report to the HTA any Serious Adverse Event (SAE) or Serious Adverse Reaction (SAR) within 24 hours of the discovery and to also notify the actions taken. It is also the responsibility of the tissue establishment to instruct third parties and end users to report to the licensed tissue establishment within 24 hours of their discovery of SAEs and SARs. A follow-up report must be provided to the HTA within 90 days which outlines the cause, and corrective and preventive actions taken to prevent recurrence. The Designated Individual at the tissue establishment is responsible for reporting to the HTA via the HTA's online submission system.

UNITED STATES

In the United States human tissue and cells intended for human application are regulated as human cell, tissue, and cellular and tissue-based products (HCT/P). They are regulated by the Centre for Biologics Evaluation and Research (CBER) which is the centre within the Food and Drug Administration (FDA) that regulates biological products for human use, among others. FDA is an agency within the US government's Department of Health and Human Services (HSS). CBER regulates HCT/Ps under 21 CFR Part 1270 (Code of Regulation, Food and Drug Administration title 21 CFR Part 1270) and Part 1271 of Human Cells, Tissues, Cellular and Tissue-based Products (Code of Regulation, Food and Drug Administration title 21 CFR Part 1271). Examples are bone, skin, corneas, ligaments, tendons, dura mater, heart valve, haematopoietic stem/progenitor cells derived from peripheral and cord blood, oocytes and semen. Parts 1270 and 1271 require tissue establishments to screen and test donors, to prepare and follow written procedures to prevent the spread of communicable disease, and to maintain records.

Part 1271 is made up of six sub-parts:

a. General provisions pertaining to the scope and purpose of Part 1271, as well as definitions
b. Registration and listing procedures
c. Provisions for the screening and testing of donors to determine their eligibility
d. Current Good Tissue Practice (cGTP) requirements
e. Certain labelling and reporting requirements
f. Inspection and enforcement provisions.

FDA published three final rules to broaden the scope of products and to introduce more comprehensive requirements to prevent disease transmission.

The first final rule requires tissue establishments to register and list their HCT/Ps with FDA (Interim Final Rule: Human Cells, Tissues, and Cellular and Tissue-Based Products, Establishment Registration and Listing; 27 January 2004). All tissue establishments engaged in manufacturer (as defined in 21 CFR 1271) of an HCT/P must register with and submit to FDA, a list of each human tissue product manufactured, within five days of beginning operations. CBER maintains a listing of currently registered HCT/P establishments at https://www.accessdata.fda.gov/scripts/cber/CFAppsPub/tiss/index.cfm.

The second final rule requires tissue establishments to evaluate donors through screening and testing (Final Rule: Eligibility Determination for Donors of Human Cells, Tissues, and Cellular and Tissue-Based Products; 25 May 2004). HCT/P establishments must screen and test HCT/P donors for communicable disease including those associated with xenotransplantation, and the procedures must be maintained in compliance with relevant subparts of 21 CFR 1271. Donor eligibility determination must be based on donor screening and testing, and certain records including a summary of records used to make donor eligibility determination must accompany the HCT/P at all times. An HCT/P must be kept in quarantine and clearly identified until the completion of donor eligibility determination.

The third final rule (Final Rule: Current Good Tissue Practices for Human Cells, Tissues, and Cellular and Tissue-Based Product Establishments; Inspection and Enforcement, 24 November 2004) establishes current good tissue practice for HCT/Ps. HCT/P establishments must follow current Good Tissue Practices (cGTPs) to ensure HCT/Ps are not contaminated, and do not become contaminated during manufacturing. The core cGTPs requirements in 21 CFR 1271 relate to facilities, environmental controls, equipment, supplies and reagents, recovery, processing and process controls, labelling controls, storage, receipt, pre-distribution shipment, and distribution of an HCT/P, and donor eligibility determinations, donor screening, and donor testing.

On May 25, 2005 FDA published an interim final rule to revise certain regulations regarding the screening and testing of HCT/P donors and related labelling (Interim Final Rule; 25 May 2005). The cGTP and other regulations are contained in 21 CFR Part 1271, along with provisions relating to establishment registration which will apply to HCT/Ps recovered on or after the rule's effective date, 25 May 2005. The frequency of routine biennial inspections is not statutory at present and is at the discretion of each district to decide based on a number of factors.

Under the FDA's regulation 21 CFR 1271.3(d)(1) and Section 361 of the Public Health Service (PHS) Act, Sections 361 and 368 (Public Health Service Act 42 USC 264 and 271), products meeting all the criteria in 21 CFR 1271 are regulated as "361 products". The criteria are

a. Minimally manipulated;
b. Intended for autologous use only reflected by the labelling, advertising, or other indications of the manufacturer's objective intent;
c. Not combined with another article, (except for water, crystalloids, or sterilising, preserving, or storage agent, if the addition of agent does not raise additional safety concerns with respect to the HCT/P); and
d. Either:

    i. Do not have a systemic effect and are not dependent upon the metabolic activity of living cells for the primary function;

or

ii. Have a systemic effect or are dependent upon the metabolic activity of the other cells for the primary function, and;

a. Are for autologous use;
b. Are for allogeneic use in a first or second degree relative;

or

c. Are for reproductive use.

The 361 HCT/Ps products are bone (including demineralised bone matrix (DBM)), skin, ligaments, tendons, fascia, cartilage, ocular tissue (cornea and sclera), arteries and veins (except umbilical cord veins), pericardium, amniotic membrane, dura mater, heart valve, hematopoietic stem/progenitor cells derived from peripheral and cord blood, semen, oocytes and embryos. These products do not require pre-market approval. Combination products such as DBM mixed with carriers such as glycerol, sodium hyaluronate, calcium sulphate and gelatine are regulated as "devices" by Centre for Devices and Radiological Health (CDRH). Bone-suture-tendon allografts are also regulated as "devices". They are subject to regulations specific to medical devices, in addition to applicable sections of Part 1271. These products also require 501(k) market authorisation.

Reporting Requirements

Part 1271 requires HCT/P establishments to report adverse reaction involving communicable disease related to an HCT/P to FDA. Each report is submitted to CBER within 15 calendar days of initial receipt of the information. HCT/P establishments must investigate all adverse reactions as soon as possible and a follow-up report must be submitted within 15 days of the receipt of new information or as requested by FDA, or the reasons why additional information is not obtainable and the steps taken to seek additional information. HCT/P establishments are also required to investigate and submit all HCT/P deviations related to a distributed HCT/P for which the establishment performed a manufacturing step. The report must be made within 45 days of the discovery of the event to CBER.

INTERNATIONAL ATOMIC ENERGY AGENCY (IAEA) CODE OF PRACTICE FOR RADIATION STERILISATION OF TISSUE ALLOGRAFTS

Since the late 1980s, the IAEA has been supporting tissue banking, especially in the developing countries, to set up and operate tissue banks using radiation as a mean to sterilise tissue (Phillips and Morales, 2003; Morales and Phillips, 2009). Gamma irradiation is the most effective method of terminal sterilisation as gamma rays provide a strong penetration power. It is cost effective, well established and widely used in the healthcare products industry. It is now routinely used for tissue sterilisation by many tissue banks and a large number of tissue grafts sterilised by irradiation has been supplied and used in human application (Myint *et al.*, 2013).

The disadvantages of gamma irradiation are adverse changes in biomechanical properties due to degradation of collagen (Nguyen *et al.*, 2007; Kamiński *et al.*, 2009) and reduction of bone inductive property as a result of adverse effect on growth factors (Munting *et al.*, 1988; Han *et al.*, 2008). For those reasons irradiation was not widely used previously, especially by the tissue banks in the United States which were employing aseptic method of processing without any terminal sterilisation. This practice was popular at that time until there was a series of disease transmissions from donor to recipient reported and confirmed (MMWR, 2002; Kwong *et al.*, 2005). Irradiation is now used by many tissue banks around the world and is the method of choice for sterilising tissue. With this change of practice there is also a development towards using a lower sterilisation dose than 25kGy which is a historical dose used in the medical device industry (Nyugen *et al.*, 2011; Myint *et al.*, 2013). It was felt that a lower sterilisation dose was necessary to reduce or minimise the adverse effect of irradiation whilst achieving the same level of sterility which is 10^{-6} sterility assurance level (SAL).

Unlike other healthcare products, the irradiation sterilisation of tissue grafts poses a particular challenge in that tissue allografts are not products which are produced in large number of samples per batch. The bioburden in tissues is much greater in type as well as in level than those found in healthcare products where the levels of microbial contamination are usually low and relatively uniform in type and level. For these reasons, extra attention must be given to uniformity of sample shape and density, bioburden in the sample, donor screening and viral contamination and the number of samples available for dose-setting purposes. A high-level international meeting on radiation sterilisation of tissue was held in the United Kingdom in 2004 and subsequently the IAEA published the technical document on the validation and routine control of radiation sterilisation of tissue allografts (IAEA, 2007). This IAEA Code of Practice provides the necessary guidance in the use of ionising radiation to sterilise tissue allografts in order to ensure their safe clinical use. The document specifies the requirements for not only radiation sterilisation of tissue allografts but also for validation, process control and routine monitoring of donor selection, tissue processing, preservation and storage. The principles adopted in the Code of Practice in terms of statistical approaches to establishing sterilisation doses are similar to those described for healthcare products; however, there are some differences in practice arising from the physical and biological characteristics of tissues.

The methods to determine the sterilisation dose are described in ISO11137 Sterilisation of healthcare products (ISO 11137-2, 2013). The inactivation of microorganisms follows the exponential law that the probability of a microorganism surviving can be calculated if the number and the type of microorganisms are known, and their resistance to irradiation. Two methods are used to establish the radiation sterilisation dose.

Method I relies on knowing the bioburden before irradiation and uses the data to establish a verification dose which indicates the dose required to deliver a SAL 10^{-2}. It uses a statistical approach to set the sterilisation dose based on three batches and therefore requires a large number of samples for initial bioburden determination and verification dose experiments. There is also an adaptation of this method for a single production batch.

Method II relies on post-irradiation burden after giving a series of incremental doses to the samples, these doses being well below the dose required for a SAL of 10^{-6}. This method needs 280 samples to determine the dose required to deliver SAL 10^{-2} before extrapolation is made to give the dose for SAL 10^{-6}. No assumption is made about the distribution of microorganisms and their resistances.

Method I has also been adapted to allow the use of as little as ten samples to determine the verification dose that gives a SAL of 10^{-1} to establish the dose required for a SAL value of 10^{-6} which is used to substantiate the 25 kGy dose. Another method of substantiating the 25 kGy dose is also available.

Within the Code of Practice it addresses two main areas of validation: pre-sterilisation processes and sterilisation processes.

Pre-Sterilisation Validation Requirements

1. Qualification of the tissue bank facility which requires the tissue bank to have appropriate facilities fit for the purpose of receiving, processing and storing tissue as defined by various tissue banking professional associations. It also requires having a documented system to ensure the standards are met and maintained, especially with regard to microbial contamination. The tissue bank should have access to an accredited microbiological laboratory to measure the level of bioburden at various levels of processing.
2. Donor selection which includes a full donor exclusion criteria based on medical and social history, age of donor, time and condition of tissue retrieval, physical examination, post-mortem results where applicable, and serology testing as required by respective rules and regulations. With regard to viral contamination, the same principal of eliminating bacterial infection applies to achieving the sterility assurance level. Irradiation can inactive the viruses; however, it should be noted that viruses have higher D_{10} value than bacteria.
3. Tissue processing and preservation which require various steps in processing depending on the tissue being processed. Examples are removal of bone marrow from bone, defatting and decontamination with chemicals. The preservation methods are freeze drying, deep freezing, heat drying and chemical preservatives.
4. Maintenance of validation such as audit, quality control, environmental monitoring, routine monitoring of microbial contamination level at

pre- and post-processing and preservation stages, and validation of processing methods and preservation.

5. Process specification which includes a quality system detailing standard operating procedures, records, product specifications, deviation, corrective and prevent actions, and other procedures which cover all types of tissue products.

All these requirements are designed to achieve a low bioburden level ensuring that microbial contamination is maintained within a set level before tissue allografts are subject to the sterilisation process.

Sterilisation Process Validation Requirements

1. Qualification of tissue allografts for sterilisation

   1.1. Evaluation of the tissue allograft and packaging

   It requires establishing a maximum acceptable dose for tissue, taking into account the adverse effect of irradiation on the physical and biological properties of tissue allografts. Likewise the choice of packaging material shall also be considered.

   1.2. Sterilisation dose selection

   It recommends establishing the number and type of microorganisms that are normally found on tissue. The determination of the sterilisation dose is based on both factors and comparison shall be made with the reference microbial resistance distribution which is adopted for healthcare products in order to set the sterilisation dose.

   1.3. Technical requirements

   This includes having access to radiation source, microbiological and dosimetric laboratory services, and tests performed using approved methods.

   1.4. Transfer of sterilisation dose

   The conditions of transferring the procedures for sterilisation dose from the source used for validation and that used for routine production are the same as in ISO 11137 Part 1. (ISO 11137-1, 2006).

2. Qualification of the irradiation facility

   The principles covering the documentation of the irradiation system and its testing, calibration and dose mapping are described in ISO standard 11137 Part 1, and apply equally to tissue allografts.

3. Qualification of the irradiation process

   3.1. Determination of the product loading pattern

   The same ISO standard 11137 Part 1 applies to tissue allograft.

   3.2. Product dose mapping

   The above standard applies to tissue; however, the tissue allografts by nature are likely to be heterogeneous and more variable in physical characteristics, especially the distribution of density within the product. These factors must be considered to ensure the actual absorbed dose can be delivered accurately. A reference shall be made to that of the verification dose experiment.

4. Maintenance of validation

   The guidelines covering calibration of equipment and dosimetric systems, irradiator requalification and sterilisation dose auditing are the same as given in ISO standards and apply equally to tissue allografts.

5. Routine sterilisation process control

   The guidelines covering process specification, tissue allograft handling and packing in the irradiator, and sterilisation process documentation are similar to those given in ISO standards and apply equally to tissue allografts.

6. Quality, safety and clinical application of the tissue allograft

   A programme to demonstrate the quality, safety and clinical application of the tissue allograft throughout its shelf life shall be performed.

Sampling procedures appropriate to the tissue type should be devised for this purpose.

7. Documentation and certification procedures

   Records related to qualification and validation of all procedures and practices such as tissue bank facility, tissue allograft, processing, preservation and sterilisation shall be reviewed by designated individuals, and the records shall be archived.

8. Management and control

   Documented control of the procedures involved in the donor selection, tissue processing and preservation prior to sterilisation and the radiation sterilisation process shall be in place.

Establishing a Sterilisation Dose

In the annexes the Code of Practice describes the practices and procedures for determining the bioburden levels of tissue allografts and the application of this information to establish the radiation sterilisation dose. Examples are given for sterilisation of amnion and bone. This is particularly useful for those in tissue banking that do not have the necessary background in irradiation sterilisation. The first step is to establish a sample item portion (SIP) for each tissue type, ensuring the SIPs are representative of not only in shape, size and composition but also in bioburden. At least 20 SIPS should be used (ten for bioburden testing and ten for verification dose experiments). The next step is to determine the bioburden which is a count on aerobic bacteria, spores, yeasts, moulds and anaerobic bacteria. The bioburden gives the total number of viable organisms within or on a tissue allograft and the packaging just before the sterilisation step. The validation of the bioburden estimation requires determination of the effectiveness and reproducibility of the test method.

Once the bioburden is established, the next step is to determine the verification dose. There are statistical calculations involved and the concept of establishing a verification dose for a SAL of 10^{-6} is to use a much higher value for SAL 10^{-2}. For this experiment samples should be taken from production batches and irradiated at the calculated verification dose. The number of samples required depends on the batch size and the number of batches. For example, a total of 20 samples is required for a minimum production batch size of 20–79. The number of positive sterility tests allowed in the verification dose experiments should be calculated statistically using an acceptable range of values of probability for a number of positive tests for sterility. For example, for ten samples taken from a batch of 20, up to one positive test of sterility is proposed, For 30 or more, up to two positive tests of sterility are proposed. There are also alternative sampling strategies available that include, for example, a double sampling plan that can minimise sample sizes and yet offer similar protection.

There are a number of methods that may be used to set the dose. The Code of Practice proposes the methods which are based on statistical approaches used previously for the sterilisation of healthcare products and modified appropriately for the typically low number of tissue samples available. The tissue bank may choose to substantiate a sterilisation dose of 25 kGy for microbial levels up to 1,000 cfu per unit or a specific sterilisation dose may be validated depending on the bioburden levels and the radiation resistances (D_{10} values) of the constituent microorganisms. Two methods are given for each approach depending on the bioburden levels and the number of samples per batch. Once the verification dose experiment is successful, the dose required to produce a SAL of 10^{-6} for the whole allograft product should be calculated using the values in the tables given in the annexes. For a 25 kGy sterilisation dose, the verification dose experiment substantiates the sterility.

The sterilisation dose calculated by the above methods can now be put into routine use if the selection of tissue and tissue processing procedures can demonstrate the consistent levels of bioburden. The level of variation in bioburden should be consistent with the sterilisation dose to be used routinely. In such cases, a sterilisation dose audit should be carried out at regular intervals of at least every three months.

Since the publication of the IAEA Code of Practice, an updated new edition in Spanish is being printed.

REFERENCES

Code of Regulation, Food and Drug Administration Title 21 CFR Part 1270 Human Tissue Intended for Transplantation (62 FR 40444 29 July 1977).

Code of Regulation, Food and Drug Administration Title 21 CFR Part 1271 Human Cells, Tissues, Cellular and Tissue-Based Products (66 FR 5466, 19 January 2001; Title 21 Vol. 8 Part 1271).

Department of Health (February 2001). A Code of Practice for Tissue Banks Providing Tissues of Human Origin for Therapeutic Purposes.

Directions given under the Human Tissue Act 2004 implementing the Human Tissue (Quality and Safety for Human Application) Regulations 2007. (November 2010). Ref 003/2010. Available at http://www.hta.gov.uk/_db/_documents/2010-11-12_General_Directions_November_2010.pdf.

Directive 2004/23/EC of the European Parliament and of the Council of 31 March 2004 on setting standards of quality and safety for the donation, procurement, testing, processing, preservation and distribution of human tissues and cells. Official Journal (OJ) of the European Union, 7.4.2004; L 102/48–58.

Directive 2006/17/EC of 8 February 2006 implementing Directive 2004/23/EC of the European Parliament and of the Council as regards certain technical requirements for the donation, procurement and testing of human tissues and cells. Official Journal of the European Union 9.2.2006; L38/40.

Directive 2006/86/EC of 24 October 2006 implementing Directive 2004/23/EC of the European Parliament and of the Council as regards traceability requirements, notification of serious reactions and events and certain technical requirements for the coding, processing, preservation, storage and distribution of human tissues and cells. Official Journal of the European Union 25.10.2006; L294/37.

Directive 2012/39/EU of 26 November 2012 amending Directive 2006/17/EC as regards certain technical requirements for the testing of human tissues and cells. Official Journal of the European Union 27.11.2012; L327/24.

Final Rule: Eligibility Determination for Donors of Human Cells, Tissues, and Cellular and Tissue-Based Products (25 May 2004) [69 FR 29785].

Final Rule: Current Good Tissue Practices for Human Cells, Tissues, and Cellular and Tissue-Based Product Establishments; Inspection and Enforcement (24 November 2004) [69 FR 68611].

Han B, Yang Z, Nimni M (2008) Effects of gamma irradiation on oseoinduction associated with demineralised bone matrix, *J Orthop Res* **26**(1), 75–82.

Human Application: HTA standards. Available at http://www.hta.gov.uk/_db/_documents/Human_Tissue_(Quality_and_Safety_for_Human_Application)_Regulations_2007_Standards.pdf.

Human Tissue Act 2004 (November 2004) The Stationery Office.

HTA Code of Practice 1, Consent (September 2009) Available at http://www.hta.gov.uk/legislationpoliciesandcodesofpractice/codesofpractice/code1consent.cfm.

HTA Code of Practice 5, Disposal of Human Tissue (September 2009) Available at http://www.hta.gov.uk/legislationpoliciesandcodesofpractice/codesofpractice/code5disposal.cfm.

HTA Code of Practice 8, Import and Export of Human Bodies, Body Parts and Tissue (September 2009) Available at http://www.hta.gov.uk/legislationpoliciesandcodesofpractice/codesofpractice/code8importandexport.cfm.

HTA Preparation Process Dossier: A Guide for Processors of Tissues and Cells for Patient Treatment (Human application). Available at http://www.hta.gov.uk/_db/_documents/Preparation_Process_Dossiers_-_A_guide.pdf.

Human Tissue 2007 No. 1523; The Human Tissue (Quality and Safety for Human Application) Regulations 2007 (May 2007) Available at http://www.hta.gov.uk/_db/_documents/Q&S_Human_Application_Regs_2007.pdf.

Human Tissue Authority (November 2010) Guide to Quality and Safety Assurance for Human Tissues and Cells for Patient Treatment. Available at http://www.hta.gov.uk/_db/_documents/Annex_-_Guide_to_Quality_and_Safety_Assurance_for_Tissues_and_Cells_for_Patient_Treatment.pdf.

IAEA (2007) Radiation Sterilisation of Tissue Allografts: Requirements for Validation and Routine Control: A Code of Practice. STI/PUN/1307, IAEA, Vienna, Austria.

Interim Final Rule: Human Cells, Tissues, and Cellular and Tissue-Based Products, Establishment Registration and Listing (27 January 2004) [69 FR 3823].

Interim Final Rule: Human Cells, Tissues, and Cellular and Tissue-Based Products; Donor Screening and Testing, and Related Labelling (25 May 2005) [70 FR 29949].

ISO 11137-1 (2006) Sterilisation of Health Care Products — Radiation, Part 1: Requirements for Development and Routine Control of a Sterilisation Process for Medical Devices.

ISO 11137-2 (2013) Sterilisation of Health Care Products — Radiation, Part 2: Establishing the Sterilisation Dose.

Kamiński A, Gut G, Marowska J, Lada-Kozlowska M, Biwejnis W, Zasacka M (2009) Mechanical properties of radiation-sterilised human bone-tendon-bone grafts preserved by different methods, *Cell Tissue Bank* **10**(3), 215–219.

Kwong FNK, Ibrahim T, Power RA (2005) Incidence of infection with the use of non-irradiated morcellised allograft bone washed at the time of revision arthroplasty of the hip, *J Bone Joint Surg Br* **87**(11), 1524–1526.

MMWR (2002) CDC update: Allograft associated bacterial infections — United States **51**(10), 207–210.
Morales J, Phillips GO (2009) The impact of the International Atomic Energy Agency (IAEA) programme on radiation and tissue banking in Asia and the Pacific and the Latin American regions, *Cell Tissue Bank* **10**, 79–86.
Munting E, Wilmart JF, Wiijne A, Hennebert P, Delloye C (1988) Effect of sterilisation on osteoinduction: Comparison of five methods in demineralised rate bone, *Acta Orthop Scand* **59**(1), 34–38.
Myint P, Wondergem J, Pynda Y, Phillips GO (2013) An international survey of tissue banking: A preliminary report, *Cell Tissue Bank* **14**(2), 159–165.
Nguyen H, Morgan DA, Forwood MR (2007) Sterilisation of allograft bone: Effect of gamma irradiation on allograft biology and biomechanics, *Cell Tissue Bank* **8**(2), 93–105.
Nguyen H, Morgan DA, Forwood MR (2011) Validation of 11 kGy as a radiation sterilisation dose for frozen bone allograft, *J Arthroplasty* **26**(2), 303–308.
Phillips GO, Morales J (2003) The International Atomic Energy Agency (IAEA) programme in radiation and tissue banking: Past, present and future, *Cell Tissue Bank* **4**, 69–76.
Public Health Service Act, Title 42: The Public Health and Welfare; Chapter 6A: Public Health Service. Cite: 42USC264 and Cite: 42USC271.
The Royal Liverpool Children's Inquiry Report: Full Text (30 January 2001) The Stationery Office, London.
The Report of the Public Inquiry into Children's Heart Surgery at the Bristol Royal Infirmary 1984–1995: Learning from Bristol (July 2001) The Stationery Office.

3

THE LEGAL AND REGULATORY FRAMEWORK FOR TISSUES AND CELLS IN THE EUROPEAN UNION

Deborah R. Hutchison

ESTABLISHMENT OF A EUROPEAN UNION (EU) REGULATORY FRAMEWORK FOR TISSUES AND CELLS

The growing use of human tissues and cells in therapeutic treatment and the development of cross-border trade make it necessary to define a minimum regulatory basis that applies across the European Union (EU). While most member states already had legislation to control the ethical aspects of donor protection, mainly in the area of organ transplantation, many did not agree on rules covering quality, safety, or the transplantation of tissues and cells.[1] In 1998, the European Group on Ethics in Science and New Technologies in the European Commission confirmed that there was "the urgent need to regulate the conditions under which human tissues circulate within the European Market."[2]

In response, the European Parliament and Council issued the European Union Tissue and Cells Directives (EUTCD) to establish a unified approach to the regulation of tissues and cells as well as high quality and safety standards regarding the procurement, testing, processing, storage, and distribution of tissues and cells. The directives introduced safety and quality standards while facilitating the safer and easier exchange of tissues and cells among EU member states. They also require that systems be put in place to ensure that all tissues and cells used in human application are traceable from donor to patient.

The importance of the EUTCD was captured well by Markos Kyprianou, European Commissioner for Health and Consumer Protection: "Hundreds of thousands of EU citizens every year undergo some form of treatment based on the use of human tissues and cells, and they must be able to trust in the safety and quality of this process. The legislation that we have set down will ensure that this is the case — wherever the tissue and cells come from or are received within the EU."

Since 1999, the EU has had the mandate, under the Amsterdam Treaty (Article 152),[3] to pass laws on the quality and safety of human tissues and cells, human organs, and blood used in medical treatment. In 2002, the European Commission put forward a framework legislative proposal to set binding requirements for the safety and quality of human tissues and cells from patient to donor.[4] The European Parliament and Council adopted this legislation on March 31, 2004 and published it in the official Journal of the European Union on April 7, 2004. EU member states have been required to implement and comply with its provisions since April 7, 2006.

Once the general framework of principles and common rules for the safety and quality of tissues and cells was established, detailed implementing directives needed to be developed to give these general provisions effect and ensure that they were applied in a harmonized way throughout the EU. In order to ensure that these implementing directives were effective, a process of consultation and consensus-building with EU government officials and tissues and cells safety experts was required. To gather all necessary input and feedback and to be able to draw on the latest scientific and technical advice required time. This explains why the more detailed, technical directives were not adopted until 2006.

THE EUROPEAN TISSUE AND CELLS DIRECTIVES (EUTCD)

The EUTCD comprises three directives: the Parent Directive (2004/23/EC), which provides the framework legislation; and two Technical Directives (2006/17/EC and 2006/86/EC), which detail requirements.

- The Parent Directive, or the European Tissues and Cells Directive, (2004/23/EC) sets quality and safety standards for the donation, procurement, testing, processing, preservation, storage, and distribution of human tissues and cells.[5]
- The first Technical Directive (Directive 2006/17/EC)[6] covers the specific requirements for the donation, procurement and testing of human tissues and cells and was most recently amended by Commission Directive 2012/39/EU.[7] The Commission adopted this directive on February 8, 2006 and all member states were required to transpose it in national law by November 1, 2006.
- The second Technical Directive (Directive 2006/86/EC) covers detailed standards for traceability, notification of serious adverse reactions and events, and requirements for processing, preservation, storage, and distribution.[8]

The directives cover all human tissues except for tissues and cells used as autologous grafts within the same surgical procedure, blood and blood components as defined by Directive 2002/98/EC, and organs and parts of organs if the parts are to be used for the same purpose as an entire organ (Article 2 of Directive 2004/23/EC). The directives apply to manufactured products derived from human tissues and cells intended for human use. For products made industrially from tissues and cells, the directive applies only to donation, procurement, and testing.

Directive 2004/23/EC[9]

Directive 2004/23/EC defines procedures for setting quality and safety standards for the procurement or donation, testing, processing, preservation, storage, and distribution of human cells and tissues. This directive requires that member states supervise the procurement of human cells and tissues, and oversee the accreditation, designation, or authorization of establishments that prepare and process tissues and cells (Articles 5 and 6 Directive 2004/23/EC). To do so, member states should designate a National Competent Authority (NCA) or Authorities responsible for implementing the requirements of the Directives (Article 4 Directive 2004/23/EC). NCAs also are responsible for inspecting establishments, verifying that appropriate control measures are in place for the procurement or donation process, ensuring that human tissues and cells are traceable from the donor to the recipient and vice versa, and establishing a donor identification system that assigns a unique code, called a Single European Code, to each donation (Articles 7 and 8 Directive 2004/23/EC).

Member states must ensure that all imports and exports of human cells from and to third countries comply with the directive's safety and quality requirements and are undertaken by accredited, designated, and authorized tissue establishments to ensure the traceability of tissues and cells (Article 9 Directive 2004/23/EC).

In addition, tissue establishments must maintain a record of their activities, which must be reported to the NCA and be available to the public (Article 10 Directive 2004/23/EC). In turn, the NCA must establish a national registry of facilities, specifying the activities for which they have been accredited (Article 10 Directive 2004/23/EC).

Member states must ensure that a system is in place to report, investigate, register, and transmit information about serious adverse events and reactions (Article 11 Directive 2004/23/EC).

This directive also addresses the selection and evaluation process for prospective donors. Member states are required to have procedures in place with respect to donor selection and evaluation, including consent, data protection, and confidentiality, and the procurement or donation process (Articles 12–15 Directive 2004/23/EC).

Member states are also required to ensure that establishments have a quality management system, designate a responsible person to ensure compliance with this directive, and confirm that personnel are qualified to perform their specified activities (Articles 16–18 Directive 2004/23/EC). Additionally, this directive addresses requirements for receipt, processing, storage conditions, labeling, packaging, and distribution of tissues and cells, as well as the documentation of these activities (Articles 19–24 Directive 2004/23/EC).

Directive 2006/17/EC[10]

The 2006 directive includes specific requirements applicable to donation, procurement, and testing of human cells and tissues. This directive requires that procurement be carried out by persons who have completed a successful training program and that establishments have standard operating procedures for verifying the donor's identity, obtaining consent or authorization, assessing donor eligibility based on specified selection criteria, and assessing required laboratory tests. To maximize the safety of the procurement or donation of human tissues and cells, this directive also requires that these activities be performed in a facility that minimizes bacterial or other contamination and be undertaken in an environment that ensures the donor's health, safety, and privacy, and requires that qualified sterile instruments and other devices be used.

This directive also establishes conditions for the selection of tissue and cell donors, living or deceased. These include exclusion criteria, for example, for people with HIV, Hepatitis, or Variant Creutzfeldt–Jakob disease. Before donations are accepted, obligatory tests and thorough medical history checks must be completed. A unique identifier code must be given to each donor to ensure proper identification and traceability of donated material.

Directive 2006/86/EC[11]

This directive includes detailed requirements for the accreditation, designation, authorization, or licensing of tissue establishments (Annex I and Annex II). Member states are further required to ensure that procurement organizations have procedures in place to ensure proper record retention of procured tissues and cells and to provide notification of any serious adverse reactions and serious events (Articles 5 and 6). To ensure traceability of tissues and cells, tissue establishments must have effective and accurate systems in place to uniquely identify and label tissues and cells using a single European identifying code (Articles 9 and 10).

Implementation of the EU Directives

EU regulations, directives, and decisions are legally binding. While regulations are directly applicable to all EU member states, directives must be adopted into national law. In doing so, a member state may change the format and content of the directive but must achieve the intended results. A directive further sets a time limit within which a member state must implement its requirements. For the EUTCD, these deadlines were as follows:

- Directive 2004/23/EC implemented by April 7, 2006
- Directive 2006/17/EC implemented by November 1, 2006
- Directive 2006/86/EC implemented by September 1, 2007

The European Commission may take legal action against a member state who either fails to implement the directive within the set deadline or implements the directive incorrectly. For example, in 2009 in the case *Commission v. Belgium* (*Case C-8/09*) the Court of Justice of the EU declared that Belgium failed its obligations to implement Directive 2006/17/EC in a timely manner.[12] Shortly after this verdict, Belgium adopted a Royal Decree in September of 2009 to implement Directive 2006/17/EC and Directive 2006/86/EC.

The EUTCD is interpreted and implemented through national authorities in each individual member state. Depending on the national interpretation of the text, the directives may be implemented differently across EU member states. The provisions set forth in a directive are considered to be minimum requirements, but member states may choose to have more stringent requirements when adopting a directive into national law.[13] For example, 14 member states apply additional testing requirements for donations, which take into account their specific national epidemiological situations.[14]

Directive 2004/23/EC specifically highlights that member states may introduce requirements related to voluntary unpaid donation, including the prohibition or restriction of imports of human tissues and cells. Additionally, member states may decide to prohibit the donation, procurement, testing, processing, preservation, storage, distribution, or use of any specific type of human tissues or cells from any specified source.[15] Furthermore, national implementation is subject to national regulations that may be already in place in each member state.

TRANSPOSITION OF EUTCD IN THE EU MEMBER STATES

While the directives were supposed to have been implemented in 2006 and 2007, the transposition of the EUTCD took longer than expected. By 2008 only two-thirds of the member states had transposed the EUTCD into national law, prompting the European Commission to encourage member states to accelerate the transposition process.[16] In May 2009, competent authorities were requested to complete a questionnaire indicating whether the three directives had been transposed into their national law. The European Commission published the results in December 2009 and indicated that the majority of member states had transposed the parent and two technical directives into national law.[17]

In June 2010, the European Commission published an updated report that provided a detailed overview of the situation in the then 27 member states.[18] This report indicated that

- All member states had designated a competent authority;
- Most member states had an accreditation, licensing system for tissue establishments in place;
- More than half of the member states imported tissues and cells from third countries and about half of them also exported tissues and cells;
- Almost all member states had a system in place to report, investigate, register, and transmit information about serious adverse events and reactions;
- Most member states embedded the principle of voluntary and unpaid donation in their national law;
- About half of member states included the conditions for disclosure of the identity of a recipient and/or donor in their national law;
- Almost all member states complied with the minimum testing requirements and some even added testing requirements;
- Tissue establishments in most member states had third-party agreements;
- Only France, Poland, and the United Kingdom reported that they would impose penalties on tissue establishments for infringements of national provisions.

Today, there are 28 member states. Croatia was the last member state to join on July 1, 2013. All member states have laws and regulations in place that set national standards for tissue and cell procurement, processing, application, and distribution. The key laws and regulations for each member state are summarized below. Please note that this is not meant to be an exhaustive discussion of all applicable laws and regulations for each member state, but the objective is to provide an overview and starting point.

Austria

The Human Tissue Safety Act from 1999, as amended in 2008 (Gewebesicherheitsgezetz BGBI I 2008/49) implements European Directive 2004/23/EC.[19] It regulates the procurement of human cells and tissues for human application as well as their processing, storage, and distribution, so long as they are not used for producing proprietary medicinal products, investigational medicinal products, or medical devices. If human cells and tissues are not intended for human application, but for example are to be used for research, they do not fall under the act. The Human Tissue Safety Act does not differentiate among the different types of cells. The act does not apply to tissues and cells used as autologous transplants within the same medical procedure, nor to blood and blood components (these are regulated by the Blood and Safety Act, Federal Law Gazette 1999/44, as last amended by Federal Law Gazette 2009/107).

Three ordinances were issued later in 2008 to implement Directive 2006/17/EC and Directive 2006/86/EC respectively, namely Gewebevigilanzverordnung 2008/190,[20] Gewebeentnamheverordnung 2008/191[21] and Gewebbankenverordnung 2008/192.[22]

The Austrian Federal Office for Safety and Healthcare is the competent authority. It is part of the Ministry of Health.

Belgium

The Organ and Tissue Law of June 13, 1986 (Belgisch Staatsblad 14/02/1987) is the first Belgian law regarding organ and tissue use and is focused mainly on organ donation. The law of December 19, 2008 on the acquisition and use of human biological material for the purpose of medical human applications or scientific research implements the EU directives.[23] The law has been revised several times since its issuance on June 16, 2009, namely December 23, 2009; July 3, 2012; and most recently on March 19, 2013. This law applies to the donation, removal, receipt, testing, modification, preservation, distribution, and use of human tissue for therapeutic and research purposes. The King and Minister Council have published regulations to further implement this law and the EU directives, such as the Royal Decree of September 28, 2009[24] to establish detailed quality and safety standards, which was amended in April of 2014.[25]

The Competent Authority is the Federal Agency for Medicines and Health Products (FAMHP) (Federaal Agentschap voor Geneesmiddelen en Gezondheidsproducten (FAGG)), which is an agency under the Ministry of Health.

Bulgaria

The Organ, Tissue and Cell Transplantation Act, first promulgated in the State Gazette issue 83/19.09.2003, has been in force since January 1, 2004 and was last amended on February 15, 2003.[26] The Act regulates the conditions and commitment for the transplantation of organs, tissues, and cells in human medicine. It applies to the collection, processing, labeling, preservation, and provision of human tissue. Also important to note is that, under the Act, by-products derived after processing organs and tissues may be used for manufacturing pharmaceuticals and medicinal products (Art 15). Regulations on the labeling, registration, and reporting of the manufacture, grafting, handling, processing, and labeling of organs, tissues. and cells may be found in the Medical Ordinance No. 22 of May 3, 2007.[27]

The Minister of Health established a legal entity, the Executive Agency of Transplantation, as the competent authority.

Croatia

In Croatia, the field of tissue and cell transplants is regulated by the Act of Conditions for Removal and Transplantation of Human Body Parts for

Therapeutic Purposes (Official Gazette 177/04) and other decrees. In 2012, Parliament issued the Law on the Transplantation of Human Organs for Therapeutic Purposes (Official Gazette 144/12)[28] and the Law on the Application of Human Tissues and Cells (Official Gazette 144/12).[29] The latter does not apply to reproductive tissues and cells, tissues and cells of the embryo and fetus, blood and blood products, tissues used as autologous graft within the same surgical procedure, organs or parts of organs if it is in the human body for the same purpose, as well as the entire organ and tissue collected during organ transplant surgery for the same organs. The Law on the Application of Human Tissues and Cells implements the EU directives regarding tissues and cells (Directive 2004/23/EC; Directive 2006/17/EZ; Directive 2006/86/EZ) and the Commission Decision (2010/453/EU).

The Ministry of Health and Social Welfare is the competent authority for tissue and cells. The Department for Health Inspection performs inspections and supervision of the blood and tissue establishments.

Cyprus

The EU directives regarding human tissues and cells have been transposed to Cypriot legislation via Law 187(I) 2007, titled Standards for Quality and Safety of Donation, Procurement, Testing, Processing, Maintenance, Storage and Distribution of Human tissues, Cells and Related Products.[30] The law applies to human tissues and cells, including hematopoietic stem cells derived from bone marrow, peripheral blood and cord blood intended for human application, and derivative products from human tissues and cells or intended for human application.

The Medical and Public Health Services Department of the Ministry of Health serves as the competent authority in Cyprus.

Czech Republic

The EU directives regarding human tissues and cells have been implemented in Czech Republic through the Law on Donation, Procurement and Transplantation of Tissues and Organs (Transplantation Act No. 285/2002

Coll. as amended by the Act. No. 228/2005 Coll., Act. No. 129/2008 Coll., Act. No. 296/2008 Coll., Act. No. 41/2009 Coll., Act. No. 281/2009 Coll., Act. No. 375/2011 Coll. and Act. No. 44/2013 Coll.)[31]

Tissue establishments and diagnostic laboratories may only carry out activities regulated by the tissue and human cell Acts if they have received permission to do so. Providers must meet requirements of the Decree to Ensure the Quality and Safety of Human Tissues and Cells Intended for Human Applications (296/2008).[32] Additional requirements related to registration, distribution, processing, labeling, packaging, and storage can be found in the 422/2008 Decree Establishing Detailed Requirements for Ensuring the Quality and Safety of Human Tissues and Cells Intended for Human Applications.[33]

Denmark

The Danish Tissue Act of 2006 (Act No. 273 of 01/04/2006) aims to ensure high and consistent quality and safety standards for the handling of human tissues and cells in Denmark. The act was fully implemented in Denmark in 2007.[34] The provisions of the Danish Tissue Act apply only to human tissues and cells intended for therapeutic use in humans. Consequently, tissues and cells used in research are not covered by the Danish Tissue Act. The Act also does not apply to tissues and cells used as autologous graft, blood, and blood components other than stem cells or organs or parts of organs, if these are intended to be used for the same purpose as the entire organ in the human body.

Executive Order No. 753 of 03/07/2006 sets the standards to ensure quality and safety for the donation, procurement, and testing of human tissues and cells.[35] This Order specifically incorporates the EU Directive 2006/17/EC in its Annex. Executive Order No. 984 of 02/08/2007 sets forth the requirements for testing, processing, preservation, storage, distribution, import, and export of human tissues and cells.[36]

The Danish Medicines Agency is the competent authority and its permission is required before an institution may test, process, preserve, store, or distribute tissues and cells. The Agency monitors compliance with the Tissue Act and regulations.

Estonia

European regulations were transposed into national legislation by the Handling and Transplantation of Cells, Tissues and Organs Act (RT I 2008, 25, 163), which came into force on January 1, 2009.[37] This Act applies to cells, tissues, and organs that are removed from a living or deceased donor and which come into immediate contact with a human organism by way of topical or systemic transplantation or in any other manner. In addition, Regulation No. 83 of December 19, 2008 establishes rules for handling cells, tissues, and organs.[38]

Institutions that handle cells, tissues, and organs need to obtain an activity license from the competent authority, the State Agency of Medicines. The handling of cells, tissues, and organs must be done in compliance with good manufacturing practices valid within the EU.

Finland

The EU tissues and cell directives were enacted in Finland through the Act on the Medical Use of Human Organs, Tissues and Cells (101/2001),[39] revised by the Amending Law of Human Organs and Tissues for Medical Use (547/2007),[40] which entered into force on June 1, 2007. The Tissue Act is applicable to the removal, storage, and use of human organs and tissues. Directives 2006/17/EC and 2006/86/EC have been transposed into national law through Administrative Regulation 3/2007 and Decree 1302/2007. The latter sets forth traceability requirements and details operating requirements, such as the need to have appropriately trained personnel, equipment, and a documentation system.[41]

The competent authority is the Finnish Medicines Agency.

France

Comprehensive legislation and regulations may be found in the Code of Public Health (Code de la santé publique), which was last updated on October 28, 2013. The sale and use of elements and products of the human body are governed by the provisions of Book II of the Code of Public Health.[42] In addition, several Decrees and Ordinances were issued over the years, including Ordinance 2007-613, Decree 2008-968, and Decree 2008-891, all of which have been incorporated into the Code of Public Health.

Organ procurement and handling may only be done by facilities that are authorized to do so by the Biomedicine Agency. The French Health Products Safety Agency (Agence française de sécurité sanitaire des produits de santé — AFSSAPS), a public administration organization under the Minister of Health, previously was responsible for the safety, quality, and proper use of health products. On April 29, 2012, France published in its official journal that the National Agency for the Safety of Medicines and Health Products (Agence nationale de sécurité du médicament et des produits — ANSM) has replaced AFSSAPS.

Germany

The most important legal stipulations governing organs, tissues, and cells are included in the following: the Law on the Donation, Removal and Transplantation of Organs and Tissues;[43] the German Drug Act;[44] the Regulation on the Tissue Establishment Registers;[45] the Regulation on the Requirements for Quality and Safety of the Procurement of Tissues and their Transmission;[46] and the Regulation on the Production of Medicinal Products and Active Substances.[47]

The German tissue laws and regulations exceed requirements of the EU directive. The German Drug Act incorporates EU directives on tissues and cells and defines "tissue" broadly to include all tissue components that are not defined as organs.

Greece

The EU directives regarding human tissues and cells have been incorporated into the Greek legal system through Presidential Decree 26/2008 (Government Gazette 51/A/24.3.2008).[48] The Presidential Decree applies to the donation, control, coding, processing, preservation, storage, and distribution of human tissues and cells intended for human applications and of manufactured products derived from human tissues and cells intended for human applications.

Law No. 3984/2011, entitled Organ Donation and Transplantation and Other Provisions (Government Gazette 150, June 27, 2011),[49] incorporated into Greek law Directive 2010/53/EU (originally 2010/45/EU). The 2011 law has been amended by Law No. 4075/2012 of 11 April 2012.[50]

Hungary

In Hungary, the legislation related to the donation and use of organs and tissues is codified in the 1997 CLIV Health Care Act.[51] Chapter XI focuses on organ and tissue transplantation and indicates in § 202[2] that it also applies to human cells. Ministry of Health Regulation 18/1998 further implements 1997 CLIV provisions related to organs and tissues.[52]

Ireland

Ireland incorporated the EU directives in its legislation through Statutory Instrument No. 158, the European Communities Quality and Safety of Human Tissues and Cells Regulations of 2006 (Human Tissues and Cells Regulations).[53] These regulations set forth requirements relating to quality and safety for the donation, procurement, testing, processing, preservation, storage, and distribution of human tissues and cells for human application. The regulations do not apply to tissues and cells used as an autologous graft within the same surgical procedure, blood and blood components, and organs or parts of organs if their function is to be used for the same purpose as the entire organ in the human body. Certain exemptions apply, including the storage of tissues and cells at sites of human application whereby a tissue establishment authorization is not required.

Ireland also incorporated European Communities (Human Tissues and Cells Traceability Requirements, Notification of Serious Adverse Reactions and Events and Certain Technical Requirements) Regulations 2007 into Statutory Instrument No. 598 of 2007.[54]

The Irish Medicines Board is the competent authority for the purpose of implementing EU and national regulations relating to human tissues and cells.

Italy

Italy has a general law dealing with the removal and transplantation of organs and tissues. This is the law of April 1, 1999, No. 91,[55] which came into force on April 16, 1999. Italy further transposed the EU directives into its legislation through several decrees. The Legislative Decree of November 6, 2007, No. 191, implements Directive 2004/23/EC on setting standards of quality and safety for the donation, procurement, testing, processing, preservation, storage, and distribution of human tissues and cells.[56] The Legislative Decree of January 25, 2010, No. 16, transposed Directives 2006/17/EC and 2006/86/EC implementing Directive 2004/23/EC with regard to the technical requirements for the donation, procurement, and testing of tissues and cells, as well as with regard to the requirements of traceability and notification of serious adverse reactions and events.[57] Decree No. 16 also applies to the procurement and testing of tissues and human cells as well as the coding, processing, preservation, storage, and distribution of human cells and tissues. It entered into force on May 3, 2010, and was amended in 2012 by the Legislative Decree of May 30, 2012, No. 85, which entered into force on November 7, 2012.[58] Finally, Directive 2003/32/EC relating to medical devices manufactured utilizing tissues of animal origin was transposed into Italian law by Legislative Decree of April 6, 2005, No. 67.[59]

Latvia

In Latvia, the 1992 Law on the Protection of the Body of Deceased Human Beings and the Use of Human Tissues and Organs in Medicine, as last amended in 2012, describes the process by which tissues and organs of living or deceased human beings may be used for scientific research and study purposes, transplantation, manufacture of medicinal preparations, and bioprosthesis.[60] This law further sets forth requirements related to pathological-anatomical and forensic examinations, the removal of tissues and organs from deceased human beings, the extraction of tissues and organs from a living donor, protection of the body of deceased humans, and inadmissibility of commercial transactions. Interestingly, the Law also highlights that if there are international or cross-border agreements, as well as conventions to which the Republic of Latvia is a member state, that contain provisions other than what is included in Latvian Law, then the international legal norms will be applied.

On October 22, 2013, tissue and organ regulations were adopted on the use of human tissues and cells by the Cabinet Nr. 1176 and published in the *Latvian Journal*. The regulations entered into force on December 11, 2013.[61]

Lithuania

The Law on Donation and Transplantation of Human Tissues, Cells and Organs from November 19, 1996 (No I-1626), which was last amended on October 19, 2006, establishes the conditions and procedures for the donation, procurement, testing, processing, preservation, storage, distribution, and transplantation of human tissues, cells, and organs and of manufactured products derived from human tissues and cells intended for human applications.[62] The provisions of the law have been harmonized with the EU Directive 2004/23/EC on setting standards of quality and safety for the donation, procurement, testing, processing, preservation, storage, and distribution of human tissues and cells.

On January 4, 2008, the Minister of Health issued Order No. V-7 on Human Tissue and Organ Donation, Procurement, Testing, Processing, Preservation, Storage and Distribution Services (Official Gazette, 2008, No. 7-251). The Order entered into force on February 1, 2011.[63]

The National Transplant Bureau is the national competent authority and was established by the Ministry of Health of the Republic of Lithuania in 1996.

Luxembourg

The 2004 EC regulations have been incorporated into the Aggregated Health Code chapter on human tissues and cells through the law of August 1, 2007 regarding human tissues and cells destined to be used in humans (Mem. A — 150 of August, 20 2007, p. 2750; doc. Parl 5448; dir. 2004/23/CE).[64] This law establishes the quality and security norms for human tissues and cells to ensure the public's health is protected. The law is applicable to the donation, receipt, control, transformation, conservation, storage, and distribution of tissues and cells, as well as products developed from human tissues and cells. On August 30, 2007, implementing regulations regarding the traceability requirement, the requirement to notify when adverse events occur, and certain technical requirements related to coding, transformation, conservation, storage, and distribution of human tissues and cells was issued (Mem. A — 169 of September 7, 2007, p. 3264; dir. 2006/86/CE; Rectificatif: Mem. A — 185 of October 9, 2007, p. 3404).[65]

Malta

The EU directives concerning the regulation of tissue banks is transposed in Maltese law by (1) Act IV of 2006 on regulating the collection and testing of human blood and blood components and establishing standards of quality and safety for human tissues and cells intended for human transplants;[66] (2) the 2006 Tissues and Cells (Quality and Safety) Regulations;[67] and (3) the 2007 Human Tissues and Cells (Coding, Processing, Preservation, Storage and Distribution) Regulations.[68]

The Netherlands

The regulatory framework regarding human tissues and cells in the Netherlands consists of the Law of February 6, 2003 on the safety and quality of human materials that can be used in a clinical setting,[69] which is complemented by the Ministerial Decree of 2006 regarding the safety and quality of tissues and cells.[70] The Law is not applicable to tissues and cells that are removed from a person and, during the same surgery, reused for the same person, nor to blood and materials that are covered by the Law on Drugs.

On August 22, 2012, the Minister of Health issued regulation GMT-3125861, which identifies the EU directives referenced in the Law on Safety and Quality of Tissues and Cells. Article 1 of the 2012 regulation indicates that the EU directive mentioned in Article 1(1)(n) of the Law on Safety and Quality of Tissues and Cells refers to EU Directive 2004/23/EG. Article 2 of the 2012 regulation indicates that Article 1(1)(o) of the

Law refers to the EU Directive 2010/53/EU (PbEU 2010, L207). These 2012 regulations came into force on August 27, 2012.[71]

Poland

The Cell, Tissue and Organ Recovery, Storage and Transplantation Act of July 1, 2005 (Official Journal of Acts Dziennik Ustaw 05.169.1411), as amended by the Act of July 27, 2009,[72] sets forth the rules of the recovery, storage, transplantation, testing, processing, storage, and distribution of human cells and tissues.[73]

In addition to the Cell and Tissue Act, a number of implementing regulations were issued by the Minister of Health, including

- The Regulation of April 2, 2010 on the Export [from] Polish Territory and Import into the Territory of Human Cells, Tissues and Organs (Official Gazette of May 5, 2010 No. 75, Item 485).[74]
- The Regulation on the requirements to be a candidate to donate cells, tissue, or organs (Official Gazette of May 10, 2006 No. 79, Item 556).[75]
- The Regulation of November 20, 2006 which sets forth technical and sanitary requirements for tissue and cell facilities (Official Gazette of November 30, 2006 No. 218, Item 1598).[76]
- The Regulation on the detailed requirements to be satisfied by the quality assurance system for tissue and cells (Official Gazette of October 24, 2008 No. 190, Item 1169).[77]
- The Regulation on the detailed conditions of collection, storage, and transplantation of cells, tissues, and organs (Official Gazette of December 16, 2009 No. 213, Item 1656).[78]
- The Regulation of March 12, 2010 (Official Gazette of April 6, 2010 No. 54, Item 331), which details the way potential recipients' qualifications are assessed.[79]
- The Regulation of March 12, 2010 on how entities that collect, store, and transplant cells, tissues, and organs are controlled and inspected.[80]
- The Regulation of March 22, 2010 details the way costs are determined for activities related to the collection, storage, processing, sterilization, and distribution of cells, tissues, and organs (Official Gazette April 2, 2010 No. 53, Item 319).[81] This regulation was later on amended by the

Regulation of October 19, 2012 (Official Gazette of November 6, 2012, Item 1215), which went into effect on January 1, 2014.[82]

- The Regulation on the training of persons whose activities directly affect the quality of cells, tissues, or organs and the safety of recipients (Official Gazette of April 20, 2010 No. 64, Item 403).[83]
- The Regulation on the unique marking and monitoring of cells, tissues, and organs (Official Gazette of May 5, 2010 No. 75, Item 486).[84]

Portugal

Law No. 22 of 2007 transposed Directive 2004/23/EC into national legislation.[85] Law No. 12 of 2009 establishes the legal regime concerning the quality and safety of the donation, procurement, testing, processing, preservation, storage, distribution, and application of tissues and cells of human origin.[86] This law transposes into national legislation the Directives 2004/23/EC, 2006/17/EC, and 2006/86/EC.

The Competent Authority is the Authority Services Blood and Transplant (ASST). ASST aims to ensure the quality and safety of tissues and cells during donation, procurement and testing, processing, storage, and distribution, including import and export.

The Portuguese Blood and Transplant Institute (IPST) is a public institute charged with coordinating the donation, procurement, processing, preservation, storage, and distribution of tissues and cells at the national level. Decree-Law No. 39/2012 of February 16, 2012 approves the constitution of the IPST and sets forth its mission and tasks.[87]

Ordinance No. 357/2008 from the Ministry of Health regulates the national network coordinating procurement and transplantation.[88]

Romania

Law No. 95/2006, Title VI focuses on the removal and transplantation of organs, tissues, and cells of human origin for therapeutic purposes.[89] Order No. 1763 of October 12, 2007 (published in Official Gazette No. 698 of October 16, 2007) transposes EU Directives 17/2006/CE, 23/2004/CE, and 86/2006/CE.[90] More detailed criteria are included in Order nr. 860/2013, including the types of professionals and rooms facilities should have, environmental and working conditions, procedures for biological samples, organizational criteria, etc.[91] Order No. 1763 includes in Annex V a sample form that lists the criteria that need to be met for a tissue bank to be accredited. This includes criteria related to the organization and management, personnel, equipment and materials, documentation, and quality assurance.[92]

The National Agency for Transplantation, Romania's Competent Authority, is a specialized public body charged with coordinating the collection, preparation, storage, validation, allocation, and transport of tissues and cells for therapeutic use in Romania. It was established and is governed by Law No. 588/2004 approving government ordinance no. 79/2004 for the establishment of the National Transplant Agency (published in Official Gazette, Part I No. 1232 of December 2, 2004) and the Regulation from March 10, 2005 (published in Official Gazette, Part I No. 228 of March 18, 2005).[93] The National Transplant Agency shall (1) promote the donation and transplantation of organs, tissues, and cells under EU directives and recommendations; (2) establish and manage the National Transplant Register; (3) develop and submit for approval to the Ministry of Health activities to be carried out in the national organ transplant programs, tissues and cells; (4) propose draft laws to the Ministry of Health; (5) propose to the Ministry of Health banks for accreditation; (6) oversee and coordinates every action of procurement; (7) work with national authorities to secure the safety of products; and (8) authorize the import and export of human organs, tissues, and cells for therapeutic use, in accordance with EU directives and recommendations.

Slovakia

Act 567/2004 on healthcare and healthcare-related services and related regulations on the execution of this Act contain rules concerning the establishment of tissue and cell banks.[94] In addition to the Health Care Act, several regulations were issued on the implementation and execution of the Act, such as Regulations 20/2007 and 622/2007.

Government Regulation No. 20/2007 Coll. (Published in the Collection of Laws, 12/2007, p. 63; effective January 14, 2007) adopts 2004 and 2006

EU regulations and regulates the details of donations of tissues and cells, criteria for donors of tissues and cells, required laboratory tests, and procedures for collection of cells and tissues.[95]

Government Regulation 622/2007 Coll. (published in the Collection of Laws 256/2007, p. 4476; effective January 1, 2008) regulates the processing, safekeeping, storage, and distribution of tissues and cells as well as the reporting of adverse events.[96] Government Regulation 41/2013 establishes the Competent Authority, the National Transplant Organization.

Slovenia

The Act on the Quality and Safety of Human Tissues and Cells Intended for Treatment (Official Gazette of RS, No. 61/2007) implements 2004 and 2006 EU directives on human tissues and cells.[97] This Act applies to the donation, procurement, testing, processing, preservation, storage, allocation, and distribution of human tissues and cells. Implementing regulations of this Act include

- Rules on the conditions for granting authorization to perform activities of human tissue and cells (Official Gazette of RS, No. 70/2008).[98]
- Rules on the acceptance, processing, storage, release, and distribution of human tissues and cells (Official Gazette of RS, No. 70/2008).[99]
- Regulations on histovigilance (Official Gazette of RS, No. 70/2008) includes more details on the system and methods of notification and reporting of serious adverse reactions and adverse events.[100]
- Rules on the method of keeping records of donations and transplantation of human body parts (Official Gazette of RS, No. 70/2003).[101]
- Rules on the conditions and procedure for import and export of human tissues and cells provide more detailed requirements regarding import and export of tissues and cells.[102]
- Rules on integration with related foreign and international organizations and exchange of human body parts with other countries (Official Gazette of RS, No. 70/2003).[103]
- Rules on the storage and transport of human body parts for transplantation (Official Gazette of RS, No. 70/2003).[104]

- Rules on the protection of personal data of donors and recipients of human body parts for medical reasons (Official Gazette of RS, No. 75/2003).[105]
- The Law on the Removal and Transplantation of Human Body Parts for Medical Reasons (ZOPDCT) (Official Gazette of RS, No. 12/2000) regulates the removal of body parts (organs and tissues).[106]

Spain

Royal Decree 1301/2006 of November 10, 2006 lays down the quality and safety standards for donation, procurement, testing, processing, preservation, storage, and distribution of cells and human tissue.[107] This Decree incorporates the 2004 and 2006 EU directives into national legislation.

Royal Decree 1825/2009 of November 27, 2009 established Spain's Competent Authority, the National Transplant Organization.[108] In general, the National Transplant Organization is responsible for coordinating transplantations of organs, tissues, and cells, as well as ensuring compliance with the regulations and quality and safety requirements for organs, tissues, and cells. Specifics on the functions and responsibilities of the National Transplant Organization can be found in Chapter II of the Royal Decree 1825/2009. The National Transplant Center and Regenerative Medicine Center was established by Royal Decree 176/2004, of January 30, 2004.[109] It is an interregional coordinating body, for which details regarding functions and structure are included in Royal Decree 167/2004.[110]

Sweden

The Act (2008: 286) on the Quality and Safety Standards in the Handling of Human Tissues and Cells transposes the 2004 and 2006 EU directives into national law.[111] The Tissues and Cells Act is supplemented by Ordinance (2008: 414) on the Quality and Safety Standards in the Handling of Human Tissues and Cells.[112] The Medicinal Products Act (2008: 12) on the management of human tissues and cells for drug manufacturing also supplements the Tissues and Cells Act and includes regulations that apply to the handling of tissues and cells used as raw material in the manufacture of pharmaceuticals for humans.[113]

The United Kingdom

The EU directives were integrated into national law through the Human Tissue Quality and Safety Regulations of 2007 and Directions given under the Human Tissue Act of 2004.[114] The Human Tissue Regulations of 2007 include the regulation of procurement, testing, processing, storage, distribution, and import/export of tissues and cells. Establishments carrying out these activities must meet the standards set forth in the Guide to Quality and Safety Assurance of Human Tissues and Cells for Patient Treatments as implemented by Human Tissue Authority (HTA) Directions 003/2010. The HTA Directions 003/2010 consolidate and clarify the standards required under the Human Tissue Regulations 2007 and revoke Directions 001/2006, 002/2007, and 004/2007.

The Human Tissue Act of 2004 covers England, Wales, and Northern Ireland. It established the Competent Authority, the HTA, to regulate activities related to the removal, storage, use, and disposal of human tissue. Scotland has separate legislation — the Human Tissue Act of 2006.[115] Both acts are based on the same principles except that the Scotland Act is based on authorization while the England, Wales, and Northern Ireland Act is based on consent.

The HTA also issued a number of Codes of Practice that came into force on September 15, 2009. The Codes of Practice provide guidance and lay down expected standards for each of the sectors they regulate. The Codes of Practice are meant to provide guidance and advice based on real-life experience. There are nine Codes of Practice, including Code of Practice 5 on disposal of human tissue and Code of Practice 8 on the import and export of human bodies, body parts, and tissue.[116]

WHAT IS NEXT?

The EUTCD was developed to establish harmonized safety and security standards deemed necessary given the increased movement of people and goods within the European Union. While member states have implemented the directives by incorporating them in their national laws and thereby have ensured a certain level of harmonization across member states, more needs to be done not only to effectively regulate import and export but also to ensure that EU citizens are able to access and trade in human tissues and cells. This is easier said than done, especially given the fact that each EU member state has legal and regulatory autonomy. While the EUTCD sets minimum standards, member states still have substantial flexibility in how these standards are implemented, interpreted, and enforced. As a result, for tissue establishments to be able to operate in the EU, they must obtain a license in each country where they want to do business and adhere to the specific national laws and regulations applicable in that country. Cross-border trade thus remains complicated despite the many efforts to ensure harmonization.

REFERENCES

1. Explanatory Memorandum, Proposal for a Directive of the European Parliament and the Council on Setting Standards of Quality and Safety for the Donation, Procurement, Testing, Processing, Storage, and Distribution of Human Tissues and Cells, available at http://eur-lex.europa.eu/legal-content/EN/ALL/?uri=CELEX:52002PC0319.
2. "Ethical Aspects of Human Tissue Banking." Opinion of the European Group on Ethics in Science and New Technologies to the European Commission. No. 11, July 21, 1998. 11p.
3. Article 152 of the Treaty Establishing the European Community, available at http://eur-lex.europa.eu/LexUriServ/LexUriServ.do?uri=CELEX:12002E152:EN:HTML.
4. Proposal for a Directive of the European Parliament and the Council on Setting Standards of Quality and Safety for the Donation, Procurement, Testing, Processing, Storage, and Distribution of Human Tissues and Cells, available at http://eur-lex.europa.eu/legal-content/EN/ALL/?uri=CELEX:52002PC0319.
5. Directive 2004/23/EC of the European Parliament and of the Council of March 31, 2004 setting standards of quality and safety for the donation, procurement, testing, processing, preservation, storage, and distribution of human tissues and cells, available at http://eur-lex.europa.eu/LexUriServ/LexUriServ.do?uri=OJ:L:2004:102:0048:0058:EN:PDF.
6. Directive 2006/17/EC of February 8, 2006 implementing Directive 2004/23/EC of the European Parliament and of the Council as regards certain technical requirements for the donation, procurement, and testing of human

tissues and cells, available at http://eur-lex.europa.eu/legal-content/EN/TXT/HTML/?uri=CELEX:32006L0017&qid=1400355834001&from=EN.

7. Commission Directive 2012/39/EU of November 26, 2012 amending Directive 2006/17/EC as regards certain technical requirements for the testing of human tissues and cells, available at http://eur-lex.europa.eu/legal-content/EN/TXT/HTML/?uri=CELEX:32012L0039&from=EN.
8. Commission Directive 2006/86/EC of October 24, 2006 implementing Directive 2004/23/EC of the European Parliament and of the Council as regards traceability requirements, notification of serious adverse reactions and events, and certain technical requirements for the coding, processing, preservation, storage, and distribution of human tissues and cells, available at http://eur-lex.europa.eu/LexUriServ/LexUriServ.do?uri=OJ:L:2006:294:0032:0050:EN:PDF.
9. Directive 2004/23/EC of the European Parliament and of the Council of March 31, 2004 on setting standards of quality and safety for the donation, procurement, testing, processing, preservation, storage, and distribution of human tissues and cells, available at http://eur-lex.europa.eu/legal-content/EN/TXT/?uri=CELEX:32004L0023.
10. Commission Directive 2006/17/EC of February 8, 2006 implementing Directive 2004/23/EC of the European Parliament and of the Council as regards certain technical requirements for the donation, procurement, and testing of human tissues and cells, available at http://eur-lex.europa.eu/legal content/EN/TXT/PDF/?uri=CELEX:32006L0017&from=EN.
11. Commission Directive 2006/86/EC of October 24, 2006 implementing Directive 2004/23/EC of the European Parliament and of the Council as regards traceability requirements, notification of serious adverse reactions and events, and certain technical requirements for the coding, processing, preservation, storage, and distribution of human tissues and cells, available at http://eur-lex.europa.eu/LexUriServ/LexUriServ.do?uri=OJ:L:2006:294:0032:0050:EN:PDF.
12. Judgment of the Court (Seventh Chamber) of September 24, 2009 — Commission v Belgium (Case C-8/09), available at http://eur-lex.europa.eu/legal-content/EN/TXT/?uri=CELEX:62009CJ0008.
13. Article 4 of Directive 2004/23/EC indicates that the Directive does not prevent a member state from maintaining or introducing more stringent or protective measures, provided that they comply with the provisions of the Treaty.
14. Communication from the Commission to the Council, the European Parliament, the European Economic and Social Committee, and the

Committee of the Regions on the application of Directive 2004/23/EC on setting standards of quality and safety for the donation, procurement, testing, processing, preservation, storage, and distribution of human tissues and cells, COM/2009/0708, available at http://eur-lex.europa.eu/legal-content/EN/ALL/?uri=CELEX:52009DC0708.

15. Article 4 (2) and (3) of Directive 2004/23/EC.
16. EUTCD Presentation by Eduardo Fernandez, EU Commission, An Overview of the European Tissues and Cells Directives: A European Commission Perspective, available at http://www.hta.gov.uk/search.cfm?FaArea1=CustomWidgets.content_view_1&cit_id=401&useCache=false.
17. European Commission, Summary Table of Responses from Competent Authorities for Tissues and Cells: Questionnaire on the Transposition and Implementation of the European Tissues and Cells Regulatory Framework, Brussels, December 2009, SANCO C6/PL/hp/D(2010)360016, available at http://ec.europa.eu/health/blood_tissues_organs/docs/tissues_responses_2008_en.pdf.
18. European Commission, Summary Table of Responses from Competent Authorities for Tissues and Cells: Questionnaire on the Transposition and Implementation of the European Tissues and Cells Regulatory Framework, Brussels, December 2009, SANCO C6/PL/hp/D(2010)360016, available at http://ec.europa.eu/health/blood_tissues_organs/docs/tissues_responses_2008_en.pdf.
19. Gewebesicherheitsgezetz BGBI I 2008/49, available at http://www.jusline.at/Gewebesicherheitsgesetz_(GSG).html.
20. Verordnung der Bundesministerin für Gesundheit, Familie und Jugend betreffend Gewebevigilanzmeldungen (Gewebevigilanzverordnung — GVVO), BGBI. II Nr. 190/2008, available at http://bmg.gv.at/cms/home/attachments/2/4/6/CH1221/CMS1359975154974/bgbl._ii_nr._190_2008__gewebevigilanzverordnung.pdf.
21. Verordnung der Bundesministerin für Gesundheit, Familie und Jugend zur Festlegung von Standards für die Gewinnung von zur Verwendung beim Menschen bestimmter menschlicher Zellen und Geweben (Gewebeentnahmeeinrichtungsverordnung — GEEVO), BGBI. II Nr. 191/2008, available at https://www.ris.bka.gv.at/Dokument.wxe?Abfrage=BgblAuth&Dokumentnummer=BGBLA_2008_II_191.
22. Verordnung der Bundesministerin für Gesundheit, Familie und Jugend, mit der nähere Regelungen für den Betrieb von Gewebebanken getroffen warden (Gewebebankenverordnung — GBVO), BGBI, II Nr. 192/2008, available at http://bmg.gv.at/cms/home/attachments/2/4/6/CH1221/

CMS1359975154974/bgbl._ii_nr._192_2008,_gewebebankenverordnung.pdf.
23. Wet van 29 December 2008 inzake het verkrijgen en het gebruik van menselijk lichaamsmateriaal met het oog op de geneeskundige toepassing op de mens of het wetenschappelijk onderzoek, available at http://www.ejustice.just.fgov.be/cgi_loi/change_lg.pl?language=nl&la=N&cn=2008121944&table_name=wet.
24. Koninklijk besluit van 28 September 2009 tot vaststelling van de kwaliteits- en veiligheidsnormen voor het doneren, wegnemen, verkrijgen, testen, bewerken, bewaren en distribueren van menselijk lichaamsmateriaal, waaraan de banken voor menselijk lichaamsmateriaal, de intermediare structuren voor menselijk lichaamsmateriaal en de productie-instellingen moeten voldoen, available at http://www.ejustice.just.fgov.be/cgi_loi/change_lg.pl?language=nl&la=N&cn=2009092808&table_name=wet.
25. Koninklijk besluit van 25 April 2014 tot wijziging van het koninklijk besluit van 28 September 2009 tot vaststelling van de kwaliteits- en veiligheidsnormen voor het doneren, wegnemen, verkrijgen, testen, bewerken, bewaren en distribueren van menselijk lichaamsmateriaal, waaraan de banken voor menselijk lichaamsmateriaal, de intermediare structuren voor menselijk lichaamsmateriaal en de productie-instellingen moeten voldoen, available at http://www.ejustice.just.fgov.be/cgi_loi/change_lg.pl?language=nl&la=N&table_name=wet&cn=2014042539.
26. Organ, Tissue and Transplation Act of 2004, available at http://bgtransplant.bg/iat/regulations.php?target_f–laws.htm.
27. Medical Ordinance No. 22 of May 3, 2007, available at http://bgtransplant.bg/iat/docs/22.pdf.
28. Law on the Transplantation of Human Organs for Therapeutic Purposes (Official Gazette 144/12), available at http://narodne-novine.nn.hr/clanci/sluzbeni/2012_12_144_3071.html.
29. Law on the Application of Human Tissues and Cells (Official Gazette 144/12), available at http://narodne-novine.nn.hr/clanci/sluzbeni/2012_12_144_3070.html.
30. The Standards for Quality and Safety (Donation, Procurement, Testing, Processing, Preservation, Storage and Distribution) of Human Tissue and Cells Act of 2007 (187 (I)/2007), available at http://www.cylaw.org/nomoi/enop/non-ind/2007_1_187/full.html.
31. Law on Donation, Procurement and Transplantation of Tissues and Organs (Transplantation Act No. 285/2002 Coll.), available at http://portal.gov.cz/

app/zakony/zakonPar.jsp?page=0&idBiblio=53599&recShow=11&fulltext=&nr=285~2F2002&part=&name=&rpp=15#parCnt.

32. Decree to Ensure the Quality and Safety of Human Tissues and Cells Intended for Human Applications (296/2008), available at http://portal.gov.cz/app/zakony/zakonPar.jsp?page=0&idBiblio=67309&recShow=2&nr=296~2F2008&rpp=15#parCnt.
33. Decree Establishing Detailed Requirements for Ensuring the Quality and Safety of Human Tissues and Cells Intended for Human Applications (422/2008), available at http://portal.gov.cz/app/zakony/zakonPar.jsp?page=1&idBiblio=67567&recShow=23&nr=422~2F2008&rpp=15#parCnt.
34. Danish Act no. 273 of 1 April 2006 on requirements for quality and safety in the handling of human tissues and cells, available at https://www.retsinformation.dk/Forms/r0710.aspx?id=10249.
35. Danish executive order no. 753 of July 3, 2006 on the quality and safety in connection with donation, procurement, and testing, available at https://www.retsinformation.dk/Forms/R0710.aspx?id=10285.
36. Executive Order No. 984 of 02/08/2007 sets forth the requirements regarding testing, processing, preservation, storage, distribution, import, and export of human tissues and cells, available at https://www.retsinformation.dk/Forms/R0710.aspx?id=11410.
37. Handling and Transplantation of Cells, Tissues and Organs Act (RT I 2008, 25, 163, available at https://www.riigiteataja.ee/en/eli/ee/RK/a/530102013104/consolide.
38. Regulation No. 83 of the Minister of Social Affairs of December 19, 2008, available at http://www.legaltext.ee/et/andmebaas/tekst.asp?loc=text&dok=XXXX034&keel=en&pg=1&ptyyp=RT&tyyp=X&query=rakkude,%20kudede%20ja%20elundite.
39. Act on the Medical Use of Human Organs, Tissues and Cells (101/2001), available at http://www.finlex.fi/en/laki/kaannokset/2001/en20010101.pdf.
40. Amending Law of Human Organs and Tissues for Medical Use (547/2007), available at http://finlex.fi/fi/laki/alkup/2007/20070547.
41. National Agency of Medicines Regulation (1302/2007), available at http://finlex.fi/fi/laki/alkup/2007/20071302?search%5Btype%5D=pika&search%5Bpika%5D=kudoslaki.
42. Book II of the Code of Public Health, available at http://www.legifrance.gouv.fr/affichCode.do?idArticle=LEGIARTI000006686056&idSectionTA=LEGISCTA000006171017&cidTexte=LEGITEXT000006072665&dateTexte=20131208.

43. Law on the Donation, Removal and Transplantation of Organs and Tissues, available at http://www.gesetze-im-internet.de/tpg/index.html.
44. German Drug Act, available at http://www.gesetze-im-internet.de/englisch_amg/englisch_amg.html.
45. Regulation on the Tissue Establishment Registers, available at http://www.gesetze-im-internet.de/tpg-gewregv/index.html.
46. Regulation on the Requirements for Quality and Safety of the Procurement of Tissues and their Transmission, available at http://www.gesetze-im-internet.de/tpg-gewv/index.html.
47. Regulation on the Production of Medicinal Products and Active Substances, available at http://www.gesetze-im-internet.de/amwhv/index.html.
48. Presidential Decree 26/2008, available at http://www.dsanet.gr/Epikairothta/Nomothesia/pd26_08.htm.
49. Law No. 3984/2011: Organ Donation and Transplantation and Other Provisions, available at http://www.eom.gr/index.php?option=com_k2&view=item&task=download&id=5&Itemid=37&lang=el.
50. Law No. 4075/2012: Organ Donation and Transplantation and Other Provisions, available at http://www.eom.gr/index.php?option=com_k2&view=item&task=download&id=15&Itemid=37&lang=e.
51. 1997 CLIV Health Care Act, available at http://www.ovsz.hu/oco/jog-szabalyok; c.
52. Ministry of Health Regulation 18/1998, available at http://net.jogtar.hu/jr/gen/hjegy_doc.cgi?docid=99800018.EUM#lbj30param.
53. Statutory Instrument No. 158. The European Communities Quality and Safety of Human Tissues and Cells Regulations of 2006, available at http://health.gov.ie/blog/statutory-instruments/the-european-communities-quality-and-safety-of-human-tissues-and-cells-regulations-2006/.
54. Statutory Instrument No. 589, The European Communities Human Tissues and Cells Traceability Requirements, Notification of Serious Adverse Reactions and Events and Certain Technical Requirements Regulations of 2007, available at http://www.dohc.ie/legislation/statutory_instruments/pdf/si20070598.pdf?direct=1.
55. Law of April 1, 1999, n. 91, available at http://www.normattiva.it/atto/caricaDettaglioAtto?atto.dataPubblicazioneGazzetta=1999-04-15&atto.codiceRedazionale=099G0153&currentPage=1.
56. Legislative Decree of November 6, 2007, n. 191, available at http://www.normattiva.it/atto/caricaDettaglioAtto?atto.dataPubblicazioneGazzetta=2007-11-09&atto.codiceRedazionale=007G0221&currentPage=1.

57. Legislative Decree of January 25, 2010, No. 16, available at http://www.normattiva.it/atto/caricaDettaglioAtto?atto.dataPubblicazioneGazzetta=2010-02-18&atto.codiceRedazionale=010G0030&currentPage=1.
58. Legislative Decree of May 30, 2012, No. 85, available at http://www.normattiva.it/atto/caricaDettaglioAtto?atto.dataPubblicazioneGazzetta=2012-06-26&atto.codiceRedazionale=012G0104.
59. Legislative Decree of April 6, 2005, No. 67, available at http://www.normattiva.it/atto/caricaDettaglioAtto?atto.dataPubblicazioneGazzetta=2005-04-28&atto.codiceRedazionale=005G0090.
60. Law on the Protection of the Body of Deceased Human Beings and the Use of Human Tissues and Organs in Medicine, available at http://likumi.lv/doc.php?id=62843.
61. Tissue and Organ Regulations of October 22, 2013, available at http://likumi.lv/doc.php?id=261810.
62. Law on Donation and Transplantation of Human Tissues, Cells and Organs from November 19, 1996 (No I-1626), available at http://www.transplantacija.lt/content/ftp/word/law.en.html.
63. Order No. V-7 on Human Tissue and Organ Donation, Procurement, Testing, Processing, Preservation, Storage and Distribution Services (Official Gazette, 2008, No. 7-251), available at http://www3.lrs.lt/pls/inter3/dokpaieska.showdoc_l?p_id=391786&p_query=&p_tr2=.
64. Law of August 1, 2007 regarding human tissues and cells destined to be used in humans (Mem. A — 150 of 20 August 2007, p. 2750; doc. Parl 5448; dir. 2004/23/CE), available at http://www.legilux.public.lu/leg/textescoordonnes/compilation/code_sante/23_TISSUS_ET_CELLULES_HUMAINS/index.html?highlight=tissus%22humains.
65. Human Tissue and Cells Regulations of 2007, available at http://www.legilux.public.lu/leg/textescoordonnes/compilation/code_sante/23_TISSUS_ET_CELLULES_HUMAINS/B_REGLEMENTS_EXECUTION.pdf.
66. Human Blood and Transplants Act (Cap. 483) to regulate the collection and testing of human blood and blood components and to establish standards of quality and safety for human tissues and cells intended for human transplants. September 15, 2006, Act IV of 2006, as amended by Legal Notice 427 of 2007, available at http://www.justiceservices.gov.mt/DownloadDocument.aspx?app=lom&itemid=8949&l=1.
67. Legal Notice 271 of 2006 Human Blood and Transplants Act, 2006 (Act No. IV of 2006) Tissues and Cells (Quality and Safety) Regulations, 2006, Government Gazette of Malta No. 17. 994-10.11.2006, available at http://

www.justiceservices.gov.mt/DownloadDocument.aspx?app=lom&itemid=12015&l=1.
68. Human Blood and Transplants Act (Cap. 483) Human Tissues and Cells (Coding, Processing, Preservation, Storage and Distribution) Regulations, 2007, Government Gazette of Malta No. 18, 141-30.10.2007, available at http://www.justiceservices.gov.mt/DownloadDocument.aspx?app=lom&itemid=12017&l=1.
69. Law of February 6, 2003, on the safety and quality of human materials that can be used in a clinical setting, available at http://wetten.overheid.nl/BWBR0014682/geldigheidsdatum_20-01-2014.
70. Ministerial Decree of 2006 regarding the safety and quality of tissues and cells, available at http://wetten.overheid.nl/BWBR0021255/geldigheidsdatum_20-01-2014.
71. Ministerial Decree of 2012 regarding the meaning of EU Directives in the Law on the Safety and Quality of Human Materials, available at http://wetten.overheid.nl/BWBR0031914/geldigheidsdatum_20-01-2014.
72. Act of July 27, 2009 on the Collection, Storage and Transplantation of Cells, Tissues and Organs, available at http://orka.sejm.gov.pl/proc6.nsf/ustawy/1657_u.htm.
73. Cell, Tissue and Organ Recovery, Storage and Transplantation Act of July 1, 2005 (Official Journal of Acts Dziennik Ustaw 05.169.1411), available at http://www.poltransplant.pl/Download/prawo/Polish_Transplantation_Act_2005.pdf.
74. Regulation of April 2, 2010 on the Export of Polish Territory and Import Into the Territory of Human Cells, Tissues and Organs (Official Gazette of May 5, 2010 no. 75, item 485), available at http://www.poltransplant.org.pl/rozp_wywoz.html.
75. Regulation of May 10, 2006 on the requirements to be a candidate to donate cells, tissue, or organs (Official Gazette of May 10, 2006 no. 79, item 556), available at http://www.poltransplant.org.pl/contents10.html.
76. Regulation of November 20, 2006 on technical and sanitary requirements for tissues and cells, available at http://www.poltransplant.org.pl/rozp_bt.html.
77. Regulation of October 24, 2008 on the requirements to be met by quality assurance systems in tissue banks, available at http://www.poltransplant.org.pl/rozp_bt_wym.html.
78. Regulation of December 16, 2009 on the detailed conditions of collection, storage, and transplantation of cells, tissues, and organs, available at http://www.poltransplant.org.pl/rozp_warunki2.html.

79. Regulation of March 12, 2010 on the operation of centers eligible for transplantation and qualifications of recipients, available at http://www.poltransplant.org.pl/rozp_ok.html.
80. Regulation of March 12, 2010 on the mode of control in entities engaged in activities related to the procurement, storage, and transplantation of cells, tissues, and organs, available at http://www.poltransplant.org.pl/rozp_kontrola_2010.html.
81. Regulation of March 22, 2010 on the detailed method of determining the cost of activities related to the procurement, storage, processing, sterilization, and distribution of cells, tissues, and organs, available at http://www.poltransplant.org.pl/rozp_koszty.html.
82. Regulation of October 19, 2012 amending the regulation on the detailed method of determining the cost of activities related to the procurement, storage, processing, sterilization, and distribution of cells, tissues, and organs, available at http://www.poltransplant.org.pl/rozp_koszty_2012.html.
83. Regulation of March 30, 2010 on the training of persons whose activities directly affect the quality of cells, tissues, or organs and the safety of recipients, available at http://www.poltransplant.org.pl/rozp_szkolenia.html.
84. Regulation of April 2, 2010 on the unique marking and monitoring of cells, tissues, and organs, available at http://www.poltransplant.org.pl/rozp_znakowanie.html.
85. Law No. 22 of 2007, available at http://www.sg.min-saude.pt/NR/rdonlyres/28F07131-EFC4-44F5-BCCD-64F73FBE5120/17518/41464150.pdf.
86. Law No. 12 of 2009, available at http://www.sg.min-saude.pt/NR/rdonlyres/6AE15430-D41D-49D7-AED9-119BADA44C90/0/0187601897.pdf.
87. Decree-Law No. 39/2012 of February 16, 2012, available at http://ipsangue.org/ipsangue2011/images/stories/Dec.Lei_n_39_2012_organica_IPST.IP.pdf ; website: http://ipsangue.org/ipsangue2011/index.php.
88. Ordinance No. 357/2008, available at http://www.sg.min-saude.pt/NR/rdonlyres/28F07131-EFC4-44F5-BCCD-64F73FBE5120/17516/0251302515.pdf.
89. Law No. 95/2006, Title VI on the removal and transplantation of organs, tissues, and cells of human origin for therapeutic purposes, available at http://www.transplant.ro/Lege/TitlulVI_Legea_95_2006.html.
90. Order No. 1290/2006, published in the Official Gazette Part I No. 916 of 10/11/2006, application of Title VI "Making sampling and transplantation of organs, tissues and cells of human origins for therapeutic purposes" of Law 95/2006 on healthcare reform, available at http://www.transplant.ro/Lege/Ordin_1290_2006.html.

91. Order No. 860/2013 accreditation criteria for transplantation or organs, tissues, and cells of human origin, Official Gazette No. 464 of July 26, 2013, available at http://www.transplant.ro/Lege/ORDIN-MS-860-2013.pdf.
92. Order No. 1763 of October 12, 2007, available at http://www.transplant.ro/Lege/ORDIN-MS-1763-2007.pdf.
93. Law No. 588/2004 approving government ordinance No. 79/2004 for the establishment of the National Transplant Agency, available at http://www.transplant.ro/Lege/Legea_588_2004.html; Regulation of March 10, 2005, available at http://www.transplant.ro/Lege/Regulament2005.html.
94. Act 567/2004 on healthcare and healthcare related services, available at http://jaspi.justice.gov.sk/jaspiw1/htm_zak/jaspiw_mini_zak_zobraz_clanok1.asp?kotva=k2&skupina=1.
95. Government regulation no. 20/2007 Coll. (Published in the Collection of Laws, 12/2007, p. 63; effective 01/15/2007), available at http://jaspi.justice.gov.sk/jaspiw1/htm_zak/jaspiw_mini_zak_zobraz_clanok1.asp?kotva=k1&skupina=1.
96. Government regulation 622/2007 Coll. (Published in the Collection of Laws 256/2007, p. 4476; effective 01/01/2008), available at http://jaspi.justice.gov.sk/jaspiw1/htm_zak/jaspiw_mini_zak_zobraz_clanok1.asp?kotva=k1&skupina=1.
97. The Act on the Quality and Safety of Human Tissues and Cells Intended for Treatment (Official Gazette of RS, No. 61/2007), available at http://www.uradni-list.si/1/objava.jsp?urlid=200761&stevilka=3297.
98. Rules on the Conditions for Granting Authorization to Perform Activities of Human Tissue and Cells (Official Gazette of RS, No. 70/2008), available at http://www.uradni-list.si/1/content?id=87587.
99. Rules on the acceptance, processing, storage, release, and distribution of human tissues and cells (Official Gazette of RS, No. 70/2008), available at http://www.uradni-list.si/1/objava.jsp?urlid=200870&stevilka=3034.
100. Regulation on Histovigilance (Official Gazette of RS, No. 70/2008), available at http://www.uradni-list.si/1/content?id=87589.
101. Rules on the method of keeping records of donations and transplantation of human body parts (Official Gazette of RS, No. 70/03), available at http://www.uradni-list.si/1/content?id=44325.
102. Rules on the Conditions and Procedure for Import and Export of Human Tissues and Cells, available at http://www.uradni-list.si/1/objava.jsp?urlid=200870&stevilka=3037.
103. Rules on the integration with related foreign and international organizations and exchange of human body parts with other countries (Official

Gazette of RS, No. 70/03), available at http://www.uradni-list.si/1/content?id=44329.
104. Rules on the storage and transport of human body parts for transplantation (Official Gazette of RS, No. 70/03), available at http://www.uradni-list.si/1/content?id=44330.
105. Rules on the protection of personal data of donors and recipients of human body parts for medical reasons (Official Gazette of RS, No. 75/03), available at http://www.uradni-list.si/1/content?id=44683.
106. Law on the removal and transplantation of human body parts for medical reasons (ZOPDCT) (Official Gazette of RS, No. 12/2000), available at http://www.uradni-list.si/1/content?id=22590.
107. Royal Decree 1301/2006 of 10 November on the quality and safety standards for donation, procurement, testing, processing, preservation, storage, and distribution of cells and human tissue, available at http://www.ont.es/infesp/Legislacin/REAL_DECRETO_NORMAS_CELULAS_Y_TEJIDOS.pdf.
108. Royal Decree 1825/2009 of 27 November, available at http://www.ont.es/infesp/Legislacin/REAL_DECRETO_ESTATUTO_ONT.pdf.
109. Royal Decree 176/2004, 30 January, available at http://www.ont.es/infesp/Legislacin/ESTATUTO_CENATMER.pdf.
110. Royal Decree 167/2004, available at http://www.ont.es/infesp/Legislacin/ESTATUTO_CENATMER.pdf.
111. Act (2008: 286) on the Quality and Safety Standards in the Handling of Human Tissues and Cells, available at http://www.riksdagen.se/sv/Dokument-Lagar/Lagar/Svenskforfattningssamling/Lag-2008286-om-kvalitets--o_sfs-2008-286/?bet=2008:286.
112. Ordinance (2008: 414) on the Quality and Safety Standards in the Handling of Human Tissues and Cells, available at http://www.riksdagen.se/sv/Dokument-Lagar/Lagar/Svenskforfattningssamling/Forordning-2008414-om-kvali_sfs-2008-414/?bet=2008:414.
113. Medicinal Products Act (2008: 12) on the management of human tissues and cells for drug manufacturing, available at http://www.lakemedelsverket.se/upload/lvfs/konsoliderade/LVFS_2008_12_konsoliderad_tom_2011_4.pdf.
114. Human Tissue Act of 2004, available at http://www.legislation.gov.uk/ukpga/2004/30/contents.
115. Human Tissue Act of 2006, available at http://www.legislation.gov.uk/asp/2006/4/contents.
116. HTA Codes of Practice, available at http://www.hta.gov.uk/legislationpoliciesandcodesofpractice/codesofpractice.cfm.

4

LITIGATION RISKS FACING TISSUE BANKS IN THE UNITED STATES

Daniel D. Harshman

Lawsuits invoke the judicial power of the state to right a wrong or seek justice from a wrongdoer. They have two purposes. One is compensatory and the other regulatory. In the first instance, lawsuits secure compensation to the injured party. Second, they encourage the defendant and others similarly situated to modify their behavior. The threat of a public trial during which one's behavior will be publicly aired, accompanied by the prospect of having to pay damages, is often incentive to alter your behavior. While there is plenty of room for debate whether lawsuits efficiently serve these purposes, there is no doubt that lawsuits have changed the business landscape in the United States. As tissue banks have grown into national and international organizations, their size has made them targets for lawsuits just like their business counterparts. The unique service provided by tissue banks makes them susceptible to particular kinds of lawsuits similar to those filed against healthcare providers. Understanding these risks and preparing to meet them requires a brief explanation of how lawyers determine what claims to bring and how lawsuits move through various stages, from filing a complaint to a getting a verdict.

Lawsuits in the United States generally follow a prescribed path of six individual stages beginning with the pleadings and culminating in a jury trial. While many cases never reach the final stage for trial, either because the parties settle or because the court dismisses the case, most parties experience the first two or three stages inclusive of discovery and dispositive motions.

PHASE 1: THE PLEADING STAGE

The pleading or motions phase begins when the complaint is filed with the court identifying the persons and entities being sued and the claims being asserted, and describing the facts that give rise to those claims. Generally, the person bringing the suit is the *plaintiff* and those being sued are the *defendants*. Defendants respond to the complaint in a written answer and may identify additional facts that may show the claim to be invalid or untimely. Though pleadings are written in a rather peculiar style, they have the practical effect of defining the legal rights and defenses that will be disputed for the duration of the case.

The plaintiff's lawyer usually has a head start during this part of the case. He or she has met and interviewed the client and reviewed the brochures, consents, or other records received by the client during his or her surgery. The lawyer has investigated public records about the potential defendants and probably obtained medical records about his client's condition before filing suit. Some lawyers even have an expert medical witness perform a preliminary review of all available documents to develop a working theory about the injury and the role that an allograft played in that injury. This investigation varies in length and depth depending on what the lawyer thinks the case is worth. The investigation done prior to the lawsuit helps the lawyer develop a theory of the case, identify potential wrongdoers, and form the foundation for the written complaint to be filed with the court. Consequently, at the start of most lawsuits, the plaintiff knows a great deal more about you than you know about his client.

PHASE 2: FACT DISCOVERY

Once the pleadings are complete, the case moves to the second stage known as *fact discovery*. This is where both sides request information and documents relevant to the claims and defenses in the pleadings. This stage is usually what everyone remembers about a lawsuit because it is the first time that witnesses or parties become personally involved in the case. This is when executives or employees must search for documents and present themselves to be cross-examined by the other side's lawyer. The discovery stage consumes the most time and often involves the greatest expense. Court rules provide each side tools to "discover" the identity of persons with knowledge of the disputed events and the facts surrounding those events. During the discovery stage, either side may serve multiple discovery requests such as a demand for oral testimony from executives or other employees, a demand for production of documents, or a demand for written responses to questions called *interrogatories*. If a party refuses to answer or attempts to evade discovery by giving irrelevant responses, the law provides other tools called *motions* to ask the judge to intervene to compel candid responses. If appropriate, the judge may order the other side to produce documents, give testimony on a particular subject, or, in extreme cases, order one side to pay the other's costs and expenses for an unjustified discovery dispute. The classic "smoking-gun" documents, the ones that contain harmful admissions, are the documents that parties fight the hardest to keep from discovery.

Fact discovery is unquestionably intrusive and intentionally so. Because an opponent may probe weaknesses in the other side's case, it is not uncommon for the rigors of discovery to force changes in the parties' positions. Where, however, both sides are resolute, the discovery phase can be time consuming for both parties. Discovery often requires exhaustive searches for historical documents which may mean searching through warehouses or electronic archives. When executives or employees testify, they must be prepared for the type of questions they will be asked, become familiar with the documents they may be shown, and be educated about the unfamiliar court procedure during which they will testify. Apart from the trial itself, fact discovery is usually the most expensive part of a lawsuit. While settlements are common, few lawsuits settle without the exchange of numerous discovery and deposition requests. As a practical matter, preparing to defend a lawsuit means preparing to meet the rigors of the discovery process.

PHASE 3: EXPERT DISCOVERY

While court rules or orders sometimes vary where the deadline appears, near the end of fact discovery, the parties must disclose expert witnesses. Expert witnesses are individuals who possess specialized knowledge beyond that commonly possessed by lay jurors. Ordinary jurors, for instance, would have no background or experience in the recovery of allografts, the transplant of allograft tissue, or the treatment of infection. If pertinent to the case, a pathologist might be called as an expert witness to explain how allografts are recovered from a donor. An orthopedic surgeon may provide expert testimony describing how allografts are transplanted and how they regenerate inside the patient after transplantation. Both sides must reveal the identity of their experts, the testimony those experts are expected to give at trial, and the evidence the experts relied on to form their opinions. The expert witnesses must write a report describing his or her qualifications, involvement in the case, and the expert opinions he or she expects to provide during trial. The expert must then give an oral deposition so that opposing parties can question the expert about his or her opinions and the research or analysis performed to arrive at those opinions.

Unlike the depositions taken during fact discovery, expert witnesses are permitted to charge for their time to give a discovery deposition. Fact witnesses, if they are subpoenaed, receive only a small, daily fee set by statute (usually less than US$20) plus mileage as reimbursement for the expense of traveling to the deposition. Experts, however, charge their customary hourly or daily fees. Since expert depositions may last an entire day (or perhaps longer), witness fees incurred during the expert discovery stage can be substantial. It is also common for both sides to retain more than one expert to testify on different issues in the case. Expert discovery does not take as long as fact discovery, but it is often far more complex and requires significant preparation to question an expert about his or her field of expertise.

PHASE 4: DISPOSITIVE MOTIONS (SUMMARY JUDGMENT)

Thus far, the stages discussed have involved conduct between the parties to the lawsuit with minimal involvement of the court. The judge does not become involved in discovery, for example, unless the parties have a dispute they cannot resolve among themselves. Dispositive motions, on the other hand, depend upon the judge's review and decision resolving a motion. Summary judgments are written requests asking the judge to throw out the other side's case in whole or in part without a trial. This process is called "summary judgment" because, if granted, it disposes of the case summarily based upon the motion papers.

The defining feature of summary judgment motions is that they are appropriate only when material facts are *un*disputed. The reason for this lies at the heart of a jury trial. Juries decide factual disputes while judges decide the law. In keeping with this dichotomy, judges do not decide factual disputes during summary judgment because that would supplant the jury's role. This "undisputed facts" requirement may seem like an odd rule for a lawsuit where opponents have an obvious disagreement or else they would not be suing one another. Fortunately, the law does not require unanimity for summary judgments. All that is required is that the facts material to the claim or defense be undisputed. Using a contract as an example will illustrate how this process works. If the parties are suing one another over a contract, they likely disagree what the contract means. However, they will probably both agree there is a contract that they each signed. In this example, the actual words of the contract are not disputed. Since there are no disputes about the terms of the contract, there are no facts that a jury must decide. The only dispute in our example is a legal question about the meaning or legal significance of the language. Thus, in our example, there is no material factual dispute involving the contract. Hence most judges would hear a summary judgment motion on that contract claim.

Unlike other phases of a lawsuit, the decision whether to bring a motion for summary judgment is entirely voluntary. Neither side is required to file such a motion. Of course, if the other side brings a motion for summary judgment, an opposition may be required because failing to oppose a motion is viewed as a concession that the motion is meritorious. Though summary judgment motions are not required, many lawyers bring them anyway. They theorize that the other side's opposition to summary judgment will reveal greater insight in the opponent's case. Even if a party loses the motion, it has gained additional information by the process and is better able to prepare for trial.

Because there is a great deal of work involved in preparing and opposing summary judgment, the mere filing of summary judgment can sometimes bring an opponent to the settlement table. More likely, however, it is the denial of summary judgment that drives settlement. If a court denies a summary judgment motion, that removes the last obstacle to the trial itself. At that point, a trial is nearly guaranteed. Once it becomes clear that trial cannot be avoided, that is frequently the time when parties get serious about their settlement options.

PHASE 5: PRETRIAL PREPARATIONS

The next-to-last phase is pre-trial preparation. Here, attorneys for each side exchange numerous documents informing one other and the court about the logistics of trial. These exchanges identify the number and identity of trial witnesses, documents to be used as trial exhibits, and whether witnesses will be presented by playing depositions that were videotaped during discovery. Parties also provide the court with all the documents the court will need to conduct the trial such as statements read during jury selection, juror questionnaires, jury instructions, and verdict forms. Since neither side can call a witness or introduce a document unless they were identified on one of these pre-trial lists, a great deal of planning must go into their preparation. Some courts also require estimates for how long each witness will be on the witness stand. Some judges place time limits for each side to present their entire case. Quite literally, the judge runs a stopwatch keeping track of the time used versus the time allotted. While this ensures that the lawyers quickly get to the point, it often requires difficult decisions and sacrifices over which witness or document to eliminate from the trial in order to meet the allotted time. Testimony, particularly cross-examination of a hostile witness, is not scripted. Even a party's own witnesses are often nervous while in the witness box and forget important details like dates or names. These temporary lapses may require extra time to calm the witness down and get back on track. Computers or projectors used to present evidence in the courtroom sometimes stop working and the lawyer must scramble to introduce the evidence in an alternative manner. In a trial, as in life, if something can go wrong, it usually does. But in a trial, these surprises eat away the time. A time estimate for trial is not a casual prediction.

From the exchange of all of this information, by the time trial arrives, each side has broad knowledge of the evidence, witnesses, and arguments available to the other side. That is, in fact, the point of the trial process. It is designed to eliminate surprise, to educate the parties about the weaknesses of their case, and the strengths of their opponent's case. This forced exchange of information encourages settlements before trial and ensures everyone will be prepared for the trial if a settlement is not reached.

PHASE 6: THE JURY TRIAL

In many cases, unless the parties consent to a trial before the judge, the trail is conducted before a jury. The point of empaneling a jury is to reach a sound and impartial decision by a jury of one's peers based purely on the evidence presented at trial. The size of a jury varies depending on the location of trial, but it normally ranges from six to 12 persons. In some ways, a lawsuit is a series of arguments at the end of which both sides write a screenplay for the jury. From the lawyer's perspective, the characters in this screenplay are chosen for you. Their script is determined by the facts and testimony recorded during discovery. The lawyer's job is to take the characters and the script as they appear, find a theme that ties it all together, and present the characters and the evidence in the most compelling and memorable way.

The jury trial phase begins with jury selection. Jurors are screened by the judge (and sometimes by the trial attorneys) to weed out anyone with an obvious bias or independent knowledge of the dispute. Ordinary citizens are drawn from the local community where the court is located. They are summoned to the courthouse to perform their civic duty as jurors to listen to and then render a verdict that resolves the case. Mostly, this is a task that jurors accept begrudgingly. When serving on a jury, individuals cannot attend to their own jobs or families. Were it not for the power of the court to require their cooperation, most jurors would never surrender their personal time to resolve a dispute between two strangers. Despite the obvious imposition, once selected to serve on the jury panel, jurors almost always take the role seriously and earnestly try to "do the right thing."

Once a jury is empaneled, the lawyers are given the opportunity to make opening statements. The opening statement explains in a non-argumentative fashion what the case is about and the evidence that will be presented, and describes what the lawyers will ask the jurors to do based on the evidence presented to them. After opening statements, the plaintiff is the first to present his or her case. After the plaintiff concludes or "rests," the defendant presents its case. Once both sides have presented all the evidence, closing arguments are made to the jury. At this point, the lawyers are given greater leeway to argue the facts and ask the jurors to draw inferences and conclusions from the evidence presented. Just as the name implies, a closing argument summarizes the evidence in an effort to persuade the jurors that the evidence requires a verdict in a party's favor. The judge instructs the jurors on the law that they should apply to decide the controversy. The jurors, however, are the fact finders. They decide which witnesses to believe and which side should prevail. The judge decides the law, but the jurors resolve any disputes about the evidence.

Jurors hear the evidence in open court, but they decide in private. Jurors are given instructions by the judge before they deliberate. The instructions contain statements of the law that apply to the dispute and a very rudimentary description of how to conduct their deliberations. The verdict is the decision the jury reaches collectively. In civil cases, whether the verdict must be unanimous depends largely on where the jury trial is conducted. In a US federal court, unless the parties agree otherwise, all jurors must agree on the verdict. Some state courts, however, allow verdicts even though the jury is not unanimous.

HOW DO LAWYERS DETERMINE WHO TO SUE AND WHAT CLAIMS TO ASSERT?

Most lawyers base their case intake decisions based on precedent. The holy grail of the legal business is legal precedent. Because the actions taken by courts are public business, judicial opinions are published and those opinion, together with statutes, rules, or regulations, represent the law on a subject. While statutes and regulations are written in broad terms to have general application, judicial opinions are more narrowly drawn to deal with specific facts and issues before the court. Hence, judicial opinions contain roadmaps describing how to bring a claim or, alternatively, how to defend one. Judicial opinions define the elements a lawyer must prove if he or she wishes to file a similar lawsuit. Under a legal doctrine known as *stare decisis*, judicial opinions are given deference by subsequent judges who strive to follow the rules established by prior cases. This process of showing deference to prior cases gives the law a measure of predictability. Absent some fundamental difference, new cases that arise involving the same subject can expect the same result. Thus, in much the same way that scientists research scientific journals, lawyers research prior decisions to locate similar cases and determine whether similar claims have been successful in the past.

Despite its allegiance to precedent, the law is not static. Over time, legislatures enact new statutes and administrators pass new regulations. Sometimes, legislatures enact new laws for the singular purpose of changing a court's previous ruling. There is an obvious tension between the concept of *stare decisis* and changes or progress in the law. Though there are exceptions, trial courts customarily defer to precedent and leave changes in the law to the appellate courts.

Before filing a case, the lawyer will frequently perform other research to learn about potential defendants. This may be something as simple as an internet search of a company or a company's website, a search of courthouse dockets to see if the company was ever sued in the past, or perhaps searches of other public records. If, for instance, a company has a patent on the processing of allografts, the lawyer may search patent records to read the disclosures about its processing methods. If the case is a personal injury matter, the patient will likely have received a brochure or pamphlet from the physician about the allograft. The lawyer will have access to those brochures. If a business is a public company in the United States, the lawyer can review the company's public filings with the Securities and Exchange Commission. A lawyer that is industrious may search social networks like LinkedIn to locate former employees that might be willing to share information about the company. When defending a client, whether it is a tissue bank or any other business, defense lawyers often perform similar searches to see what the opponent may have learned from a pre-suit investigation.

COMMON LAWSUITS FILED AGAINST TISSUE BANKS

At least in terms of legal precedent, tissue banks have grown up in the shadow of their older sibling, the blood bank. More widely known and with a longer history, blood banks have created much of the legal framework that tissue banks confront in present-day litigation. These cases serve as a natural starting point for lawyers and judges even though there are some fundamental differences between blood and tissue banks. Because of the similarities in legal issues, litigation involving blood banks, organ transplants, and clinical trials are frequently cited in tissue bank litigation.

The most frequent claims asserted against tissue banks, blood banks, and similar entities are negligence claims alleging the transmission of disease from the transfusion or implantation of infected blood or tissue. Some early but well-known cases involved the transmission of Cruetzfeld-Jacob disease from dura matter,[1] osteomyelitis and osteonecrosis from dental allografts,[2] and hepatitis C from implantation of bone allografts.[3] These claims typically allege that the tissue bank negligently screened donors, failed to recover the allografts properly, or negligently processed the allograft by either doing something wrong or failing to do something that was required. In a garden variety negligence case, the defendant's conduct is measured against a reasonable person standard which asks the jury to decide whether the defendant acted as a reasonably prudent person would have done. In many but not all jurisdictions of the United States, legislatures have enacted statutes that define tissue banks as "healthcare providers." The net result of these statutes is to transform a negligence lawsuit against a tissue bank into a claim for professional negligence similar to medical malpractice. The standard for professional negligence is more stringent. It requires proof that the tissue bank's conduct fell below that standard of skill, prudence, and care exercised by other tissue banks in similar circumstances. In a professional malpractice case, the "standard of care" is not based upon what a reasonable person would do, but rather what other professionals exercising ordinary skill of the profession would do. By definition, a professional malpractice case almost always requires testimony from an expert skilled in the profession. In the case of a tissue bank, for instance, the expert witness might be a physician who has experience in recovering allografts and transplanting them into patients. Since ordinary jurors have no experience in such matters, they must depend upon testimony from experts in the field to tell them what a reasonably skilled professional would do in similar circumstances.

To give some perspective, it helps to read how the law has been applied in other cases. The 1992 California decision involving the Irwin Memorial Blood Bank is a good example of how the professional negligence standard is applied.[4] The case arose in 1983 when a contaminated blood transfusion infected the recipient with the HIV virus. The patient's lawsuit claimed the blood bank was negligent because it would have detected the virus if it had used a new but as yet unproven test; the HIV virus would have been detected and the blood discarded, thus avoiding the transmission. The evidence at trial revealed that in 1983 when the blood bank performed the transfusion, no other blood banks were performing this test. Measured against what other skilled blood banks were doing, the blood bank's conduct was not negligent. In this instance, the blood bank satisfied the professional standard of care because the defendant used the same level of skill and attention as other blood banks in similar circumstances.

While many states have passed legislation exempting tissue banks and blood banks from product liability laws, that is not true in all states of the United States. Under traditional legal principles, a user of a product that did not purchase the product himself, such as an employee using a machine purchased by his or her employer, would find it difficult to sue the manufacturer if he or she was injured by a defect in the machine. The employee could not allege breach of contract since the employee did not purchase the machine. Likewise, traditional negligence theories presented high evidentiary hurdles since the employee would have to prove the manufacturer was negligent in some way. Given the rapid industrialization of the country and the rise in injuries from products, the law recognized a new type of claim known as strict liability or, more commonly, product liability. This is a type of legal claim that reduces the burden for persons injured by products, and who wish to sue the manufacturers or distributors of those products. It is a potent tool for plaintiffs. Lawyers soon adopted these products liability theories to assert claims against blood banks and tissue banks. In response, in many states, legislatures intervened to declare that blood and tissue banks offer "services, not products." Where these statutes apply, tissue banks may be protected from product liability lawsuits.

Industry standards and federal regulations also play significant roles in tissue bank litigation. Regulations promulgated by the US Food and Drug Administration (FDA) are often viewed as the absolute minimum for the standard of care. If a tissue bank does not comply with the FDA regulations, it cannot act with the same level of skill and prudence as other tissue banks who comply with the law. Because regulatory compliance can play a prominent role, lawsuits charging tissue banks with professional negligence often delve into whether the tissue bank's operations satisfy the FDA requirements and the good tissue practices of the American Association of Tissue Banks (AATB). A lawyer suing a tissue bank will be familiar with these rules and regulations because the expert hired to consult on the case will educate the lawyer about these standards. Consequently, written discovery requests and oral deposition testimony taken during the discovery phase will focus upon the steps taken by the tissue bank to comply with these rules and regulations.

Thus, to defend a negligence lawsuit requires a thorough investigation of the tissue bank's procedures for screening of donors and detection and prevention of disease. To address the question whether the tissue bank actually followed its own procedures during recovery, processing, and distribution of the plaintiff's particular allograft, you and your lawyer must use the tissue banks records to verify how the allograft was treated throughout each and every step before it was placed into the hands of the plaintiff's surgeon. Next, the lawyer must measure compliance with what other tissue banks do by comparing procedures to the FDA rules and the AATB guidelines. Since professional negligence cases turn on "expert" testimony, these evaluations must be done not only by the law firm internally, but by the expert hired to testify in the case. Given the changes made to processing and procedures over time, it is by no means an easy task to identify the documents and witnesses knowledgeable for each of these steps for the particular allograft implanted in the plaintiff.

Another area of frequent litigation is informed consent. Historically, an unauthorized touching of another person constituted an offense called "battery." If, however, the person consented to the touching of his or her person, there could be no offense. The "informed consent" doctrine applies these concepts to the doctor-patient relationship. Among other things, the physician must warn the patient of the risks of any surgical procedure such as the implant of an allograft. The doctor must obtain the patient's consent to the surgical procedure and be careful that the surgical procedure does not exceed the consent given. This principle also applies to donor families who must consent to the removal of allografts from remains of their deceased family members. Documentation of proper consent is an FDA requirement so it may be easy to overlook the fact that donor families have rights concerning their loved one's remains. Absent proper consent from the donor families and removal of allograft from the decedent may give rise to a lawsuit from donor family members. The most famous example of this type of lawsuit arose when a number of tissue banks unknowingly accepted tissue from recovery agencies run by Michael Mastromarino who notoriously forged paperwork, stole cadaveric tissue, and sold it without consent of donor families.[5] In this extreme example, the unlawful removal of remains was criminal conduct that harmed not only the donor families, but also the recipients who unknowingly received allografts that were otherwise ineligible for transplant. Extreme cases like the Mastromarino case can give rise to fraud or misrepresentation claims for those that knowingly make false statements about the screening, recovery, or processing procedures.

Another type of lawsuit frequently brought against tissue banks is a "failure to warn" claim. If a tissue bank knows of a particular risk with the allograft, it must provide warnings about that risk so the patient can either avoid using the allograft or use the allograft in a manner that mitigates the risk. Of course, this presupposes that scientific or medical knowledge at the time the allograft was distributed was such that the tissue bank knew or should have known about the particular danger. If so, the tissue bank must provide warnings to users about the known dangers of the allograft. This may seem like an odd requirement as the tissue bank may not even know the identity of the patient. In most settings, it is the physician or surgeon who has the patient relationship and the scientific knowledge to appreciate the warnings given. Courts recognize the realities of the physician–patient relationship and hold that a tissue bank, much like a drug manufacturer, discharges its duty to warn by providing adequate warnings to the physician. In the eyes of the law, the physician is a "learned intermediary" who must pass along to the patient any warnings received from the tissue bank.

Importantly, multiple legal theories are commonly brought side by side in the same complaint. Lawyers are trained to bring all potential claims in one case. The defense of a multi-theory case is more difficult for the obvious reason that each theory must be investigated and defended on its merits. This often requires more than one expert and a greater volume of discovery requests.

Of course, tissue banks are susceptible to the same kinds of employment, intellectual property, or contract disputes that confront any large-scale business. Tissue banks hire and fire employees, enter into contracts, obtain patents for novel processes, and engage in many business transactions. Everyday business transactions can also be the source of litigation, but they may not pose the same risk as lawsuits arising from the recovery, processing, and distribution of allografts. The number of allografts recovered, processed, and distributed annually probably outnumbers by a wide margin the number of employees hired or fired in one year. That means the likelihood of tissue bank litigation is probably far more likely than other types of litigation.

HOW TO PREPARE FOR LITIGATION IN US COURTS

While there is no magic formula to avoid lawsuits, there are steps that lessen the risk of being sued and reduce the burdens of being involved in litigation. Careful training of employees and regular audits of regulatory compliance will go a long way toward minimizing the risk of litigation. Having a good insurance program may help defray the cost of litigation. Insurance coverage is a valuable resource because most liability insurance policies provide a defense in the event of a lawsuit. The insurer hires the lawyer, pays for the experts and bears much of the financial burden of defending the claims. The tissue bank should take a hard look at the territorial limits of coverage to ensure that the policies apply to each of the geographic territories where allografts are recovered or distributed. To ensure that the tissue bank has proper coverage with adequate coverage limits, a risk department or insurance broker should be consulted to review the business liability insurance. Given the benefits of insurance, it should be one of the first items on the list.

Lawyers should carefully examine the tissue bank's literature, both internal and external, to assess whether the language used is "jury friendly." Tissue banking may be old hat to people in the industry, but jurors exposed to the process for the first time during trial may be startled to learn that cadaveric tissue is recovered and transplanted into other humans. While they soon learn the benefits of this process, it is important to explain the process using respectful language. This is so because brochures or internal operating procedures are often displayed to the jury during trial. That means disrespectful language that might be found in these documents will be paraded before the jurors throughout the duration of trial. Given that tissue banks are especially sensitive to donor families, it is surprising to see somewhat uncomfortable language in internal operating procedures. A few examples refer to the recovery of allografts from a decedent as the "harvesting" of tissue or refer to decedents as "subjects." These undoubtedly began as scientific terms, but they sound horrible. Farmers harvest corn. People do not harvest other people's body parts except in horror movies whose very purpose is to make people cringe. There is nothing inherently wrong with using scientific terminology, but you should consider whether some of the more harsh-sounding terms might someday be twisted by an opponent to suggest the defendant tissue bank does not care about donors, donor families, or recipients of its allografts. It is well worth the time and effort to scrutinize internal documents to re-evaluate the terminology employed throughout those documents.

Lawsuits are retrospective by nature. They require investigation of past events, digging deep into company archives, and producing historical documents. If allograft processing or recovery is at issue, the recovery and processing events have usually occurred years prior to the filing of the lawsuit. This means that responding to discovery may present logistical problems of locating and retrieving old records. Employees who were involved in those historical events may no longer be employed with the company. Hence, the only source of information may be the written record. Because FDA rules and AATB good tissue practices require considerable record keeping, most tissue banks assume that historical records are within easy reach, but this is not always the case. Over time, record-keeping processes change and computer networks migrate to new technology. Retrieving records from legacy systems is far from ideal and frequently an unexpected roadblock. Having records is one thing, being able to find and search them is quite another. These logistical problems come to the forefront at the outset of discovery because there are time constraints for responding to discovery. Discovery responses are due within 30 days of service of a discovery request. To prepare for the logistical burdens of discovery, a tissue bank should consider performing an internal survey of its records retention system. It should determine where its records are located, who has possession or access to them, and what form they take (paper or electronic). It should try retrieving a few specific records to see how easy or difficult it is to search those records. For electronic files, it should identify a roadmap to its computer servers or identify important evidence held by employees on isolated desktops or laptops. It should know what records (and whose emails) are stored on which server, how often they are backed up and how frequently electronic files are routinely destroyed. It should make certain that it has procedures in place to suspend the automatic or routine destruction of documents upon being notified of a lawsuit.

Particularly in personal injury cases, a tissue bank should expect to receive discovery requests for the records mandated by the FDA and AATB Good Tissue Practices. The plaintiff's lawyer will be familiar with these rules from his or her pre-suit investigation. The lawyer will serve discovery to determine the tissue bank's compliance with those requirements. Given the detailed record-keeping requirements these rules impose, the tissue bank will almost always have more documents to produce during discovery than the individual plaintiff. Experienced lawyers who handle plaintiff's cases well understand this disparity and many attempt to exploit the asymmetry by serving multiple document requests on the tissue bank. Responding to repeated document requests may seem like the "death of a thousand cuts." Though it is not supposed to be a tool of harassment, discovery can be misused and those who are practiced at it, know enough to disguise their goals. The burden of litigation, particularly those flowing from discovery, may tempt you to settle the case simply to avoid continued impositions caused by the lawsuit. If the abuse is sufficiently obvious, the tissue bank may ask the court to impose limitations on discovery. However, in most cases, having an organized and efficient record-keeping system is the most efficient way to ease the discovery burden.

Record retention often comes to the forefront of litigation when one side claims the other has destroyed relevant evidence. The law requires parties to preserve evidence potentially relevant to threatened or pending claims. This duty arises whenever a party is on notice of a potential claim so the duty may arise before a lawsuit is filed. To comply with this duty, tissue banks must notify all departments and employees known to have relevant records about the lawsuit and instruct them not to delete or destroy emails, documents, or other records. The tissue bank must suspend any document retention programs that might otherwise routinely destroy records under a pre-planned schedule. If a tissue bank does not have a process in place to comply with these preservation duties, a lawyer and IT professional should be consulted to create one.

The tissue bank's internal procedure should be reviewed for complaint handling. Tissue banks are well acquainted with the FDA's regulations on tracking, complaint files, and adverse incident reporting.[6] Plaintiff's lawyers scrutinize these systems very carefully. In some cases, the lawsuit may involve allegations that the tissue bank was negligent because its internal complaint handling systems were inefficient or defective. The FDA and AATB guidelines are fairly robust so issues rarely arise because the tissue bank failed to have procedures. Difficulties arise because the records that exist (or those that can be found) do not rise to the level one might expect given the procedures in place. On this score, if a tissue bank receives internal complaints by employees about tissue banking operations, those internal complaints should be thoroughly investigated and documented. While not every employee complaint has merit, slack attention to employee complaints can be a mistake. The last thing you want to face in a lawsuit is testimony from a former employee who contends that your tissue bank ignored a serious problem which eventually led to a lawsuit. Right or wrong, jurors tend to view former employees as whistleblowers. Testimony from former employees can be devastating if the tissue bank has no records to address the accuracy of their testimony. If an employee suggests the tissue bank has a problem, it should investigate those allegations and document the things it did to investigate, the persons who were interviewed, the tests that were performed, the outside persons who were consulted, and how the issue raised was resolved. If methods or records were changed as a result or it was concluded that no changes were appropriate, the reasoning must be documented. Also, the person conducting this investigation must be qualified to do the investigation. Someday, a jury may examine that investigation and what they see could weigh heavily on the ultimate verdict.

Lastly, in the event litigation arises, the tissue bank should appoint an individual in the organization to serve as liaison with the outside lawyer or insurance adjuster handling the case. This person should be familiar with the overall operations of the tissue bank. If he or she does not know the answer to a question, he or she must be able to find the person who does know the answer. If the tissue bank has in-house counsel or a risk management staff, they are the most likely candidates for this role. No matter how experienced the outside defense lawyer may be, he or she will not be familiar with the tissue bank's internal processes, taxonomy or records. Inevitably, the name a lawyer gives to a record or process when asking for those documents will not match the name the company gives those records. If discovery requests are responded to in the literal sense without a knowledgeable liaison to interact and guide the outside lawyer, important records or critical employees could be overlooked. Ideally, the liaison should also be someone with authority inside the company. It is all too easy for employees to disregard inquires that seem remote to their everyday workload. This rarely happens if a request comes from someone they recognize as senior management. Litigation may not be the only thing of importance inside the company, but if the company is to successfully defend or prosecute a lawsuit, it must afford that lawsuit a measure of priority. Appointing a liaison from senior management ensures that employees will respond in a timely and proper fashion. It also helps to ensure that the liaison has the skills and knowledge necessary to manage an important project for the tissue bank. Litigation is after all a large-scale project whose outcome depends not only on the merits of the dispute, but the logistical resources devoted to its management.

REFERENCES

1. Bourquin v. Melsungen, No. cv-88-0346322S, 1993 Westlaw 280304 (Conn. Super., field 7-15-93). Rev'd 40 Conn. App. 302 (Conn. App. Ct. 1996)
2. Arneson v. Michigan Tissue Bank, No cv. 05-189-M-JCL, 2007 Westlaw 4698986 (D. Mont., field 3-26-07).
3. Condos v. Musculoskeletal Transplant Foundation, 208 F. Supp. 2d 1226 (D. Utah 2002).
4. Osborn v. Irwin Memorial Bood Bank (1992) 5 Cal. App. 4th 234.
5. In Re Human Tissue Products Liability Litigation, 582 F. Supp. 644 (D.N.J. 2008).
6. 21 C.F.R. §1271.290 (tissue tracking), §1271.320 (complaint files) and §1271.350 (reporting).

5

THE IMPORTANCE OF A CODE OF ETHICS AND A CODE OF PRACTICE FOR TISSUE ESTABLISHMENTS

Jorge Morales Pedraza

The use of human tissues in certain medical practices has been accepted as an effective modality to save lives and to improve the quality of life of thousands of individuals after passing specific medical treatments.

The success rates for transplantation of human tissues have ensured that processed tissues can be used more frequently for specific medical treatments without increasing the risk of the recipient. To avoid the wrong uses of the human tissues procured by a tissue establishment, it is necessary to promote and adopt certain ethical principles as well as certain norms and regulations that should be observed by the operators and medical staff during the procurement, processing, storage, distribution and use of human tissues. The purpose of observing ethical principles and certain norms and regulations is to avoid not only the abuse and unethical practices in tissue banking, but also the violation of specific legal healthcare regulations in force in the country.

Human tissue procurement and transplantation are no longer considered as extraordinary or uncommon healthcare events. As the techniques become more refined, the possibility of a human tissue transplant becomes a hope for more and more people suffering different diseases. Because the demand for human tissues is great and the supply limited, dilemmas related to allocation of resources are eminent and unavoidable. However, all activities of procurement of human tissues as well as any other activities carried out by the tissue establishment require an authorisation and the observance of legal norms and regulations in force in the country. This authorisation must be subject to compliance with basic ethical principles and with health safety standards (which themselves are an ethical imperative), and all legal norms and regulations in force in the country in the field of tissue banking. For this reason, human tissue donation should not be practised unless there is no feasible alternative and means are in place to effectively protect the donor's health and safety. Tissue establishments have an eminent responsibility to monitor the implementation of these legal rules and regulations as well as ethical principles. They should refuse to accept human tissues the procurement of which does not satisfy ethical principles and safety rules, and which have not been procured following all legal norms and regulations in force in the country in the field of tissue banking.

There are also questions regarding the eligibility of recipients for human tissue transplant. Many concerns arise as to the determinations of potential recipients. Transplantation may involve tissues from dead or living human donors, and there are dilemmas and legal issues associated with each.

With living donors, issues related to voluntary informed consent or authorisation and the transplant of human tissues free of contamination are of great concern. There are some individuals who are deemed desperate and have prompted the sale of tissues to raise money for personal or family needs. With the urgency for a decision due to time factors for tissue procurement, coercion could be an important factor to be considered. In some settings, nurses are asked to approach patients or families about considering human tissue donation. In such situations nurses need to be clear about their own feelings regarding human tissue procurement and transplantation, and they must remember that attention to family needs takes precedence over the time constraints of tissue procurement (Bauer *et al.*, 2004).

It is important to note that current practices in human tissue transplantation raise several questions that need to be addressed jointly by clinicians, scientists, health regulators and ethicists, as well as representatives of civil society, particularly by human tissue donors and recipients. Although a number of norms and regulations on human tissue transplantation have been adopted in the past several years or are currently under discussion in several countries, national norms and regulations and oversight of human tissue transplantation as well as ethical guidance still are non-existent or are ineffective in many countries around the world. For this reason, the lack of ethical guidance and the existence of unmonitored or unregulated human tissue procurement and distribution represent a serious international risk to both donors and recipients of cells and human tissues (Schulz-Baldes *et al.*, 2007).

Finally, it is important to highlight the following: countries should be committed to providing oversight of all transplant activities, with national competent healthcare authorities ensuring legislation, regulations and standards, particularly the adoption of a code of ethics and a code of practice as well as an ethical policy and principles in the field of tissue banking. The adoption of ethical policy and principles, a code of ethics and a code of practice are essential elements in order to promote high standards in human tissue transplantation. The ethical principles and the codes of ethics and practice must deal with the issue of access to transplantation, issues of quality and safety, and a rational and cost-effective use of transplantation, among others. The role of the World Health Organization (WHO) and the International Atomic Energy Agency (IAEA) as well as of the regional professional associations already established in the field of tissue banking, can facilitate the formulation of global principles, norms and regulations in the field of tissue banking, keeping in mind the regional and cultural diversity.

MAIN ETHICAL ISSUES

Wherever tissues are removed from a human being, and possibly transplanted into another human being or to the same donor, the activities involved in the procurement and use of such tissues should be subject to ethical requirements intended to safeguard respect for the human being, their dignity and autonomy, as well as to follow certain legal norms and regulations.

The fact that one donor may yield several tissue transplant products, which could be used later on for specific medical treatments or research activities, highlights the need for safety measures in the procurement, processing, storage, distribution and use of human tissues. The impact of any misconduct during these processes may be understandably high. For this reason, each country should set itself the objective of guaranteeing to its citizens high and ethical standards in all healthcare activities, particularly in the field of tissue banking, and should adopt legal norms and regulations with the aim of ensuring that the procurement of human tissues will be carried out with the highest possible standards and with the protection of the donors and recipients in mind. This protection must extend equally to all tissue donors and recipients, as well as to all healthcare professionals and technicians who are involved in the procurement, processing, distribution and use of human tissues in certain medical treatments or in specific research activities.

It is important to highlight that all activities related to the procurement of human tissues should protect the altruistic aspects of donation and should be carried out under the authorisation of the national competent healthcare authorities in the country. This authorisation must be subject to compliance with basic ethical principles and with health safety standards, which themselves are an ethical imperative, and should respect all legal norms and regulations that the country has in force in the field of tissue banking (Morales Pedraza and Herson, 2011).

Governments should take all necessary measures to ensure that any promotion and publicity activities in support of the donation of human tissues comply with the guidelines or legislative provisions laid down by the country and/or local laws and/or norms and regulations. Such guidelines or legislative provisions should include appropriate restrictions or prohibitions on advertising the need for, or availability of, human tissues with a view to offering or seeking financial gain or comparable advantage.

THE ADOPTION OF A CODE OF ETHICS AND A CODE OF PRACTICE

While there are inescapable acts of individual judgment involved in ethical decisions, it is desirable that conclusions should be reached in ways that result in as consistent a practice as possible. A way of achieving this in a given area of activity is to formulate a code of ethics and a code of practice providing general guidelines within which individual decisions are to be made. A code of ethics is a basic presentation of principles and ideals and has a regulatory, educational and cultural integration function. A code of practice is a detailed description of the actions that are to be performed and not performed, and of the sanctions that attach to non-performance (Morales Pedraza, 2011).

A code of ethics has two aspects: (1) the content comprising the requirements, rules, principles, ideals and so forth; and (2) the commitment of the members of an organisation to conform to, and otherwise uphold, those rules and ideals. The commitment to that content on the part of the members of the organisation consists of an oath or promise or other kind of undertaking that each member has given to adhere to the code.[1] A code of ethics is a good place for members of an organisation to signal their commitment to the basic laws, and thus basic moral standards, of their community, including the principle of individual autonomy (Miller, 2002).

With the purpose of ensuring respect to the human being and the deserved dignity and autonomy,[2] each country should have in force a code of ethics and a code of practice with the purpose of applying them to all tissue banking activities carried out in the country.

The purpose of these codes is to promote a culture of openness, trust and integrity in all tissue banking activities carried out in the country, and to allow the use of human tissues processed in one country in another country, with a minimum contamination risk for the recipients of such tissues.

It is important to note that a successful implementation of a code of ethics and a code of practice requires the active participation and support of all tissue establishment personnel and medical staff associated with the activities of the tissue establishment.

---

[1] It is important to highlight the following: Ethics is not the same thing as etiquette or manners. So, in general, a code of ethics should not address issues of politeness or dress. Ethics is not the same thing as aesthetics. A code of ethics should not address issues of style, rhetoric or self-presentation. A code of ethics refers not only to moral principles, but also to ideals of the good. Sanctions — applied in the context of an appropriate complaints and discipline system — ought only to be deployed in relation to minimum standards (morality), but not in relation to ideals (ethics) (Miller, 2002).

[2] The respect for autonomy requires that those who are capable of deliberation about their personal choices should be treated with respect for their capacity for self-determination (CIOMS-WHO, 2008).

Tissue establishment staff should not tolerate any wrongdoing or improper ethical conduct at any time and by anyone and should take appropriate measures to act quickly with the purpose of correcting any deviation from the code of ethics and the code of practice adopted by the national competent healthcare authorities in the field of tissue banking. Any infractions of the code of ethics and the code of practice by some tissue establishment staff may be subject to disciplinary actions, up to and including termination of employment. In the case that the violator is the tissue establishment itself, the sanction to be applied could be up to the suspension of its operational licence.

There are several elements that should be considered during the discussions of any ethical issues related to a code of ethics and a code of practice associated with the procurement, processing, sterilisation, storage, distribution and use of human tissues. Among these elements are the following:

- All tissue establishment staff has the right to expect and the responsibility to conduct itself all the time within high ethical standards and legal principles;
- The tissue establishments should operate respecting all laws and regulations adopted by the national competent healthcare authorities in the field of tissue banking;
- No payment shall be given or received regarding tissue donations and other activities carried out by the tissue establishments, outside of those authorised by the national competent healthcare authorities;
- Conflicts of interest must be avoided between the tissue establishment personnel, and donors or the donor's family and legal representative;
- All tissue establishment personnel are encouraged to report any suspected wrongdoing by the staff during the fulfilment of their activities. However, it is expected that tissue establishment staff do their work with integrity, honesty and diligence;
- The tissue establishments should facilitate and support valuable research activities using human tissues for the benefit of the community, but within the legal and ethical framework established by the national competent healthcare authorities;

- Tissue establishment staff should provide the best possible services to tissue donors, end-users and recipients, and should place the highest priority on quality, timeliness and competitiveness in providing these services (Morales Pedraza and Herson, 2011).

Based on what have been said before, the following ethical principles should be considered during the preparation of any code of ethics to be followed by tissue establishment operators in the future:

- Be honest and adopt an ethical conduct, including ethical handling of current or apparent conflicts of interest between personal and professional relationships;
- Comply with applicable government laws, rules and regulations in force in the field of tissue banking;
- Maintain the confidentiality of information entrusted to the tissue establishment by the tissue donors or the donor's family or legal representative with regard to tissue donation and donor screening, except when an authorised or legal action obligates the tissue establishment to disclose this information to duly designated national healthcare authorities;
- Deal fairly with tissue donors, donor families, legal representatives, medical personnel, transplant coordinators, volunteers and other hospital staff, during the procurement, processing, and use of human tissues;
- Provide information that is accurate, completely objective, relevant, timely and understandable to all national competent healthcare authorities duly authorised to request this information;
- Proactively promote ethical behaviour among tissue establishment staff, with the purpose of creating a positive work environment within the tissue establishment and the tissue banking network;
- Protect and ensure the proper use of tissue establishment assets and tissue donations, carefully weighing against practices that may be linked to commercialisation or other hidden agendas not authorised by the national competent healthcare authorities (Morales Pedraza and Herson, 2011).

Other elements that can be considered during the preparation of a code of ethics are the following:

- Maintain the highest standards of professional conduct;
- Assume responsibility and accountability for individual judgments and actions;
- Gain and maintain proficiency of practice;
- Participate in efforts to establish, implement and improve standards of practice;
- Contribute to the growth of professional knowledge;
- Respect individual privacy and hold confidential all information obtained in the course of practice;
- Maintain a cooperative and respectful relationship with other transplant practitioners and healthcare professionals;
- Serve the public regardless of race, ethnicity, culture, social or economic status, personal political position and ideology or religious belief;
- Educate the public about donation and transplantation issues;
- Act to protect the public when healthcare and safety are endangered (Bauer *et al.*, 2004).

Codes of ethics typically prescribe and proscribe actions. Indeed, codes of ethics are largely about actions. However, the view that codes of ethics are simply rules that prescribe what actions are to be performed, and what actions are not to be performed, is too narrow. For one thing, codes of ethics typically mention virtues and vices, as well as actions. Virtues and vices are habits or dispositions to act. As such, they are not simply actions; they are elements of a person's character. For another thing, codes of ethics are, or ought to be, concerned with affective attitudes as well as actions. Attitudes are important; attitudes motivate actions (Miller, 2002).

Finally, it is important to remember the basic distinction that exists between legal norms and ethical norms. While the former are founded on the latter, there is no necessary one-to-one correspondence between each legal and ethical norm. A law may be regarded as unethical by some people (e.g., a law prescribing the death penalty for certain crimes) and likewise, an ethical norm may be regarded as unlawful in a country (e.g., one involving female genital mutilation). Thus, it cannot be expected that ethical guidelines, which translate ethical principles in the form of recommendations (rather than of strict norms), will always coincide with legal prescriptions. This applies all the more to international guidelines, which are issued in the context of legal dispositions varying from one country to another (CIOMS-WHO, 2008).

MAIN ETHICAL PRINCIPLES

With regard to the removal of human tissues, the main ethical principles from the donor's point of view are the following:

- Respect for the human body, even after the person's death;
- Respect for the autonomy of the donor; thus, tissue may not be removed when the person refuses.[3] For deceased persons, this implies that tissues may not be removed, if the person refused consent or do not give authorisation for the removal of tissues during his or her lifetime;
- Protection of vulnerable people, namely people unable to give consent or authorisation or have a difficult economic situation;
- Respect for private life and medical confidentiality;
- The right to prior information on the conditions of removal and the expected use of the human tissues;

[3] In order to exercise the right of autonomy, a patient must first possess the capacity to make decisions. When speaking of a patient's ability to decide, physicians are actually speaking of a patient's "capacity". A patient who is able to make medical decisions is considered to possess capacity. A patient who is not mentally or psychologically able to make medical decisions is considered to lack capacity. When assessing a patient's capacity, the physician must evaluate three distinct aspects of decision-making ability: (1) Patients must show that they understand the given information about diagnosis and treatment and that they appreciate the significance of the disease and its consequences. In testing for understanding, the physician might ask patients to rephrase the information he has given them; (2) Patients should be able to deliberate in accordance with their own values. Here, the physician might ask patients what is most important to them in making their decision; (3) Patients should demonstrate an ability to communicate consistent choices regarding their decisions. Here the physician might determine patients' choices at different times to test consistency (Davenport, 1997).

- The right not to be subjected to unfair discrimination, which could result from the revelation of data collected from the donor or the family, to third parties (e.g., employers and insurance companies).

The main principles affecting recipients of allografts are:

- Respect for the autonomy of the person concerned, which requires clear information on the risks and advantages of the proposed transplant;
- Respect for private life and medical confidentiality, which applies to all medical treatment;
- The right to safety, which requires prior verification of the quality and safety of the human tissue to be transplanted with regard to the risk of transmission of infectious diseases, neoplasms and immunological diseases;
- The right of patients to have fair access to the therapeutic possibilities offered by the transplantation of human tissues.

Finally, it is important to highlight that several countries accept, as a matter of principle, that tissue transplantation programmes be founded on the philosophy of voluntary and unpaid donation, anonymity of both donor and recipient, the altruism of the donor and the solidarity between donor and recipient. To support this philosophy, several governments are taking steps to encourage a strong public and non-profit sector involvement in the provision of human tissue transplantation services and the related research and development using human tissue.

According to Morales Pedraza (2012), the following are a group of specific ethical principles that can be used for the preparation of an ethical policy, a code of ethics and a code of practice by the national competent healthcare authority in any country interested in increasing the impact of tissue banking activities in the community:

- Ethical principles related to the informed consent or authorisation to donate human tissues;
- Informed consent or authorisation in case of research projects;
- Ethical principles associated with the use of procured human tissues for medical treatment and for research activities;

- Ethical principles related to the use of surgical human tissue residues;
- Ethical principles related to the confidentiality access of the donor and recipient information;
- Ethical principles related to the protection of health of donors and recipients;
- Ethical principles related to the human body integrity;
- Ethical principles related to the distribution of human tissues for medical use;
- Ethical principles related to the commercialisation of human tissues.

It is important to ensure that all medical and tissue banking activities are subject to ethical standards and legal regulations adopted by the national competent healthcare authorities that promote respect for all human subjects and protect their health and rights. Some research populations are particularly vulnerable and need special protection. These include those who cannot give or refuse consent for themselves and those who may be vulnerable to coercion or undue influence (WMA General Assembly, 2008). In addition, the patients should be informed about potential commercial use of their product and, for this reason, should be asked to release all interest for commercial use. Physicians should consider the ethical, legal and regulatory norms and standards involving human subjects in their own countries, as well as applicable international or regional norms and standards.

CODE OF PRACTICE

According to Morales Pedraza and Gajiwala (2012), a code of practice is an important document containing a general description of the actions that are or are not to be performed by a tissue establishment. It should be applied by all tissue establishments supplying human tissues for therapeutic or for research purposes. The Code of Practice after approval by the national competent healthcare authorities should become part of the set of legal norms and regulations in force in the country in the field of tissue banking. All tissue establishments should carry out their activities respecting the content of the approved code.

In order to provide safe human tissues of reliable quality, current good practice standards must be observed in the selection of donors, retrieval of human tissues, testing, processing, sterilisation, storage and delivery of human tissues for transplantation. A code of practice should address all relevant issues in a manner that could be used by all tissue establishments without any difficulty.

A code of practice is defined as "written guidelines issued by an official body or a professional association to its members to help them comply with its ethical standards and should specify the requirements for the activities of tissue establishments for the procurement and processing of human tissues for therapeutic use and research activities" (Business Dictionary).[4] The scope of a code of practice should include all human tissues used for therapeutic purposes, including those used in clinical trials and research activities.

The objective of a code of practice is to provide a set of principles and procedures that, when followed by tissue establishments, helps to ensure that the human tissue processed will have the required quality and can be used without affecting the recipient.[5] A basic tenet of a code of practice is that quality cannot be tested into a product, but must be built into each product during all stages of processing. A code of practice will also reduce the likelihood of risks associated with quality and safety of all processed human tissues, and to ensure timely access to these types of products for different medical practice. For regulatory effectiveness, it is imperative that the content of the approved code of practice, as well as all applicable legislative instruments and guidelines, are reviewed regularly or whenever significant concerns are raised regarding the adequacy of the content of the approved code. To ensure that legal norms and regulations in the field of tissue banking remain relevant and credible, it is important that the

---

[4] Codes of practice are prescriptive in style and may be referenced by regulations or certain conditions of licence. They contain practice-specific requirements that must be satisfied to ensure an acceptable level of safety in dealings with human tissues and the use of the ionising radiation technique for tissue sterilisation.

[5] The purpose of a code of practice is to set down the responsibilities of tissue establishment staff in regulating their work. The purpose of workforce regulation is to protect and promote the interests of tissue establishment users. A code of practice is intended to complement rather than replace or duplicate existing tissue establishment staff's policies and it forms part of the wider package of legislation, requirements and guidance that relate to the employment of staff. Employers are responsible for making sure that they meet the standards set out in the code, provide high-quality services and promote public trust and confidence in tissue establishment activities.

code of practice and the associated standards are regularly updated. Most countries revise and update their code of practice about every five years to keep up with changes in the industry and changes in manufacturing and testing technology. The marketplace for tissues has undergone great change during the past decade with advances in technology in manufacturing practices, as well as changes to the type and nature of tissues produced that need to be taken into consideration.

There are a number of ways that the objective of a code of practice can be achieved, including:

- Ensuring that there is a uniform and a minimum set of processing requirements established for all human tissues;
- Facilitating compliance with the processing and technical requirements through clearly written requirements;
- Achieving international harmonisation of processing requirements, where possible and appropriate, to better facilitate global cooperation and reduce regulatory burden;
- Ensuring that the existing processing requirements will allow for advances in technology or changes in industry practices correcting any legislative discrepancies identified with, including technical requirements in the existing code of practice.

A code of practice should be based on a quality system approach. It forms the basis of good practice in all tissue establishments and will be used when a tissue establishment is inspected by the national competent healthcare authorities of the country in which it is operating. For this reason, a code of practice should set out safety requirements with respect to processing, which includes donor screening, donor testing, collection/retrieval, preservation, packaging, labelling, quarantine, storage, record keeping, distribution, importation, error, accident and adverse reaction investigation and reporting. The objective of a code of practice is to maximise the safety of human tissues, by clearly stipulating the safety requirements adopted by the national competent healthcare authorities, thus making them mandatory[6] (Health Canada Publication, 2009).

---

[6]A code of practice should clearly distinguish between legal requirements, situations where the competent authority is insisting on something, statements of best practice and more vague aspirations. Those storing and using human tissue need to know exactly what is unlawful and exactly what is regarded as incompatible with holding an authority licence.

A code of practice sets out all the requirements for good management practice that collectively ensures that processed human tissues consistently meet specifications, and while it sets the benchmark for practices that should be followed, alternative approaches may be permitted, provided it can be demonstrated that the intent of the code is met in a timely and effective manner in order to meet quality objectives and respect ethical standards as well as all legal norms and regulations in force in the country in the field of tissue banking. It is the tissue establishment's responsibility to determine the most effective and efficient quality process.

Furthermore, it is important to note that the requirements of a code of practice are *minimum* requirements. Many tissue establishments have already implemented comprehensive, modern quality systems and risk management approaches that may well exceed the minimum standards.

A code of practice should not limit the development of new concepts or technologies, which have been validated and which provide a level of quality assurance at least equivalent to those set out in the code.

Finally, it is important to highlight that a code of practice needs to provide clarification of scientific principles and objectives that apply to emerging cellular and tissue technologies. It also needs to provide a tissue establishment with the flexibility to apply tailored approaches to manufacturing controls suitable to their product. A code of practice should also promote the harmonisation of regulatory practice at all levels. According to Indech (2000), the globalisation of today's marketplace impacts every sector of business, including the healthcare sector. Regulatory healthcare authorities and the industry have long acknowledged that as national boundaries are crossed on a regular basis, the harmonisation of healthcare legislation — resulting in an international consensus on essential safety, efficacy and quality principles — offers significant benefits to all parties, including the public. These benefits include enhanced public health protection on a global level (where the best elements of national healthcare systems are melded together); increased government efficiency, as governments rely on each other's expertise and experience; lower healthcare costs to the public; and the increased development and availability of advances in medical technology (including drug and medical device products).

REFERENCES

Bauer K, Taub S, Parsi K (2004) Ethical issues in tissue banking for research: A brief review of existing organisational policies, *Theor Med Bioeth* **25**, 113–142.

Davenport J (1997) **Ethical Principles in Clinical Practices**, *The Permanente Journal*, Portland, USA.

Ethical Principles for Medical Research Involving Human Subjects (2008) World Medical Association Declaration of Helsinki, 59th WMA General Assembly, Seoul, Republic of Korea, October 2008.

Guidance Document for Cell, Tissue and Organ Establishments: Safety of Human Cells, Tissues and Organs for Transplantation (2009) Health Canada Publication; Minister of Health; Canada.

Indech B (2000) The international harmonization of human tissue; regulation: Regulatory control over human tissue use and tissue banking in select countries and the current state of international harmonization efforts, *Food and Drug Law Journal* **55**.

International Ethical Guidelines for Epidemiological Studies (2008) Council for International Organisations of Medical Sciences (CIOMS) in collaboration with the World Health Organisation (WHO), CIOMS, Geneva, Switzerland.

Miller S (2002) Model Code of Ethical Principles, Centre for Applied Philosophy and Public Ethics, Charles Sturt University, for the Professional Standards Council of New South Wales and Western Australia, June 2002.

Morales Pedraza J (2011) A model of a code of ethics for tissue banks operating in developing countries, *Cell and Tissue Banking* DOI 10.1007/s10561-011-9279-8.

Morales Pedraza J (2012) The Use of the Ionizing Radiation Technique for Tissue Sterilization: The International Atomic Energy Agency (IAEA) Experience, Nova Science Publishers, Inc.; New York, USA.

Morales Pedraza J, Gajiwala AL (2012) A model code of practice for tissue banks, *IJOE* **8**(1–2).

Morales Pedraza J, Herson MR (2011) The importance of ethics in the field of human tissue banking, *Cell Tissue Bank* DOI 10.1007/s10561-010-9232-2.

Schulz-Baldes A, Biller Andorno N, Morgan Capron A (2007) International Perspectives on the Ethics and Regulation of Human Cell and Tissue Transplantation, Institute of Biomedical Ethics, Center for Ethics, University of Zurich, Zurich, Switzerland and Gould School of Law, University of Southern California, Los Angeles, CA, USA.

6

ETHICAL, RELIGIOUS, LEGAL, AND CULTURAL ISSUES FOR TISSUE BANKING, WITH SPECIAL REFERENCE TO THE ASIA-PACIFIC REGION

Aziz Nather and Wee Lin

INTRODUCTION

Each country has its own set of values, ethics, religious beliefs, legal considerations, and cultural issues. These values will inevitably affect the development of tissue banking in the country. This chapter discusses such common issues encountered in tissue banking legislation in the Asia-Pacific region.

ETHICAL CONSIDERATIONS

Donation of tissue is an act of humanity. It enables a person to alleviate the suffering of a fellow human being. Tissue banks should not be allowed to sell tissues for commercial profit. Rather, they should provide tissue grafts on a non-commercial basis.

However, since costs are incurred during procurement, processing, and distribution, tissue banks may incur "processing costs", provided the law in the country (if any) makes provisions to allow for the retrieval of such costs. A common acceptable practice in institutional banks is to work out the total costs of such procurement and processing — including manpower costs, equipment maintenance costs, cost of consumables, electricity and water consumption costs, etc. and charge these costs to the recipients or organization using these tissues (Nather, 2000). Some tissue banks in the Asia-Pacific region charge processing costs, for example, those in Japan, Singapore, Malaysia, Sri Lanka, and India.

Most commercial tissue banks are located in, or associated with, the United States. There are very few commercial tissue banks operating in countries within the Asia-Pacific region. Korea is the only country in this region which has commercial tissue banks. In other countries within the Asia-Pacific region, commercial banks are generally not favored. The view held by most countries in terms of developing commercial or private banks is to favor non-profit institutions, and most of them are against for-profit institutions if commercialization is to take place at all.

Tissue banking helps to reduce a country's healthcare costs (Hachiya *et al.*, 1999). Allogeneic bone grafts are less expensive than custom-made prostheses or ceramics, which can be costly. Commercially produced bone allografts are also prohibitive in costs. However, bone allografts produced by non-commercial or institutional tissue banks are often either provided free or at nominal costs ("processing costs only") and hence reduce the patient's medical costs.

As tissue banking involves the removal of tissues from a human being and its subsequent transplantation into another person, a strict standard of ethics is required to be put in place to ensure that the dignity and autonomy of human beings are not compromised. Moreover, to ensure the safety of potential tissue recipients, it is important to ensure that proper safety measures are used during tissue procurement, processing and distribution. This is important in order to maintain public trust in both the tissue banking and medical communities. Regulation of tissue banking is also important to ensure that tissues are not misused or sold for profit, and that the rich do not benefit at the expense of the poor. As such, an ethical code in tissue banking is vital to ensure that the use of tissues is properly regulated both nationwide and across different nations. This subject was discussed in more detail in Chapter 5.

In Europe, tissue banks comply with the Ethical Code of the European Association of Tissue Banks (EATB, 2003). In the United States, the American Association of Tissue Banks (AATB) provides standards to help ensure that tissue banks follow good ethical practice. The AATB emphasizes the need for voluntary donation and public education to help people make informed choices when it comes to tissue donation. It also stresses the need to respect the wishes of the individual and that of his or her family, maintain confidentiality, respect and honor the gift of tissue donations by a person, and avoid the distribution of tissues in a discriminatory manner. However, no ethical code has been produced in either the Latin America or Asia-Pacific regions. Nevertheless, tissue banks in the Asia-Pacific region comply with all the principles embodied in the Ethical Code of EATB. The guiding principle followed by all tissue bank operators is based on moral principles, human duty and proper conduct, as enshrined in the Hippocratic Oath (which requires doctors and health workers to "not injure." (*non nocere*).

RELIGIOUS CONSIDERATIONS

Tissue donation is a sensitive issue that invokes important concerns regarding the dignity of the living and the dead, the concept of brain death, and the belief that tissue donation is the greatest gift one can bestow upon a fellow human being after one's death. The answers to these questions are inextricably tied to the dominant religious and cultural mindsets within each country. Hence, religion plays a major role in promoting or retarding the development of tissue banking in each country. In the Asia-Pacific region, Buddhism, Islam, Christianity, and Hinduism are major religions.

Buddhist View on Tissue Donation

While there are no specific doctrines in Buddhism that are explicitly for or against tissue donation, a key idea in Buddhism is to show compassion for others and relieve suffering. Hence, tissue donation is often regarded as a gift of generosity. In fact, the large majority of Buddhists are strong advocates of tissue donation.

Many Buddhists honour those who donate their bodies and organs to advance medical science and to save lives. In Buddhist scriptures, there are stories where the donation of tissues has been referred to as acts of charity that earn merit. Buddhists are expected to meditate about the impermanence of life. The body will decay, just as a beautiful fragrant flower withers and decays. The concept of tissue donation is encouraged not only after death; even while living, tissue donation is considered to be a noble act.

In countries where Buddhism is the predominant religion, there is no shortage of tissue donors. These countries include Vietnam, Thailand, Sri Lanka and Myanmar. The most successful public awareness programs on tissue donation have been achieved in Sri Lanka, Thailand, and Vietnam. In Sri Lanka, the non-governmental organization led by Dr Hudson Silva, the Eye Donation Center, has provided over 60,000 corneas for corneal transplantation in 57 countries. Inspired by the success of the Eye Donation Center, a model human tissue bank was set up in Sri Lanka by the International Atomic Energy Agency (IAEA).

Buddhism is also one of the major religions in Korea (22.8% as of 2005) and Singapore (33.3% as of 2010). The success of the National University Hospital Tissue Bank in Singapore is largely due to the fact that the Buddhist community in Singapore strongly supports the tissue transplantation program. Most of the tissue donors in Singapore are Buddhists.

With that being said, the Southern tradition (Sri Lanka and Southeast Asia) in Buddhism believes that rebirth occurs immediately after death and hence, allows for autopsies upon death. However, the Northern tradition (East Asia and Tibetan region) believes that the body should not be touched or moved for eight hours after death as there is an intermediate state between reincarnations. As such, tissue donation is often still very much an individual decision.

Islamic View on Tissue Donation

Muslims are by far the most controversial religious group with regard to tissue donation. The Islamic states in the Asia-Pacific region include Malaysia, Bangladesh, Pakistan, and Brunei. In addition, Islam is the predominant religion in Indonesia, a secular country that follows the five principles of *Pancasila*. There are about 200 million Muslims in China, another secular country. Islam is also an important religion in other secular countries, such as Singapore (14.7% as of 2010) and India (14.6% as of 2011) (Nather, 2000).

The Koran respects life and values the needs of the living over those of the dead. Adherents of Islam strongly believe in saving human life. This means that organ donation and transplantation can be considered under circumstances where it would save a person's life, as long as the sanctity and respect for human life is maintained. No mention is made about allowing transplantation to improve the quality of life of a recipient. As a result, Muslims are more likely to allow kidney donation and less likely to allow tissue donation, as the latter is perceived as merely improving the quality of (rather than saving) life. Nevertheless, while interpretations of the Koran vary according to different religious leaders, e.g, between the *ustazs* and the *ulamas*, tissue donation is not explicitly forbidden in the Koran (Nather, 2000). In 1983, the MUIS (the Islamic Religious Council of Singapore) initially rejected organ donation, but it has now reversed its position.

Countries with a significant Muslim presence have their own *muftis*. A *mufti* is a religious official who is appointed by the government to deal with all Islamic matters in the country, including the issue of organ and tissue donation. *Fatwas* are religious rulings made by a *fatwa* committee as an official stand by the government on certain issues, e.g., the tissue donation and transplantation issue. The *fatwa* committee — chaired by the *mufti* — may include prominent religious leaders, doctors, lawyers, and members of the public.

A common misconception among Muslims is that organ and tissue donation is not permitted by Islamic Law. However, *fatwas* concerning organ donation have been declared in several countries in the Asia-Pacific region, including Brunei, Malaysia, and Singapore. A *fatwa* was passed in Saudi Arabia in 1985 sanctioning both the live and cadaveric donation of organs. Likewise, in 1998, a *fatwa* was passed in the United Arab Emirates that sanctioned live and cadaveric organ donation as well as organ donation from Muslims to non-Muslims, and that accepted the concept of brain death (El-Shahat, 1999).

A milestone event for *fatwas* specific to tissue donation occurred on September 4, 1995, when the first *fatwa* on bone, skin, and amnion was introduced by the Malaysian Islamic Center. This was followed on June 29, 1997, by a *fatwa* on bone, skin, and amnion in Indonesia, sanctioning tissue procurement from deceased donors. This was a great leap forward for Indonesia, as the previous law — the Indonesia 1992 Health Regulation — allowed tissue procurement only from living donors. Unfortunately, efforts by the author since 1995 to seek a similar *fatwa* for Muslims in Singapore from the *Majlis Ugama Islam Singapura* (MUIS, the Islamic Religious Council of Singapore) have not succeeded so far, although a *fatwa* for cornea was passed in 1999 — the first *fatwa* for tissues in Singapore.

However, *fatwas* are not legally binding, and so the decision to donate remains very much the prerogative of the individual and his or her family. Hence, the introduction of favorable *fatwas* is only the first step in promoting public acceptance of tissue donation among Muslims.

Another important consideration for Muslims is that they must bury the body as soon as possible after death. Therefore, procedures like tissue procurement, which may delay the burial, are not taken very kindly.

Culturally, Muslims accept that God created them whole and they prefer to return to Him whole. It is a common practice among many Muslims to bury amputated limbs, foreskins from circumcision, and amnions from delivery. Not all Muslims follow this practice as it is a cultural practice and not a religious requirement. Ultimately, the decision to donate tissues or organs still rests with the individual.

Therefore, it does not come as a surprise that there is a big shortage of bone donors in countries where Islam is the predominant religion, including Pakistan, Bangladesh, Malaysia, and Indonesia. The lack of donors also slows down the development of tissue banking in these countries. They have been successful in procuring amnion, but not bones and ligaments, from deceased donors.

In Islam, the *waris* issue is also important. When a person dies, the *waris* or "next of kin" plays a key role as to what happens to the body of the deceased. Even if the donor has consented, the next-of-kin must also consent if tissue donation is to be allowed. For Muslims, therefore, consent is required not only from the patient, but also from his or her next-of-kin or *waris*.

Nevertheless, the demand for tissues in these countries is great and steadily increasing. Indonesia and Malaysia have resorted to the production of bovine bone xenografts for bone transplantation until such time that public awareness programs can produce better results (Nather, 2000). Specific *fatwas* for bone have been passed in both countries, but they have not produced significant changes in the Muslim population's attitude toward tissue donation. More public education is needed to change entrenched cultural practices and beliefs, along with the passing of *fatwas*, before more Muslims will come forward and pledge to be tissue donors.

Christian View on Tissue Donation

Christianity is the predominant religion in the Philippines and Australia (Nather, 2000). It is also one of the major religions, though not the predominant one, in Korea (18.3% in 2005) and Singapore (18.3% in 2010). Tissue donation is considered to be consistent with the ecclesiastical Christian dogma of loving one's neighbor as oneself, as it is thought to be an act of genuine altruism — of giving something up at little or no cost to the donor to save the lives of others. This was reiterated by the late Pope John Paul II while attending the Congress of the Society for Organ Sharing on June 20, 1991, when he mentioned the words of Jesus Christ as narrated by the apostle Luke: "'Give, and it will be given to you; good measure, passed down, shaken together, running over, will be put into your lap' (Luke 6:38). We shall receive our supreme reward from God according to the genuine and effective love we have shown to your neighbor." These words are in full support of organ and tissue donation and transplantation. Christian communities in Europe and the United States support tissue transplantation. For instance, on the weekend of November 13–15, 1998, churches and synagogues across the United States encouraged their faithful to sign donor cards (*Japan Times*, 1998).

Despite the strong Christian presence in the Philippines and Korea, however, other factors (including cultural factors) have led to a shortage of donors in these countries. The cultural concept is again that God created them whole and they would like to return to Him whole, not physically altered by the act of tissue donation.

Hindu View on Tissue Donation

Hinduism is the predominant religion of India, a secular country. It is also an important religion in Sri Lanka (about 12.6% in 2011), Singapore (about 5.1% in 2000), and Malaysia (7.1%).

Hinduism is parallel to Buddhism in many ways: both religions practice cremation of the body, which is in fact an act of destruction of the body, in front of and with full knowledge of the relatives. Hindus are not prohibited by religious law from donating their organs and it has no objection to tissue donation and transplantation. Donation is an individual decision, but *dharma* ("good duty") suggests that doing good deeds for others is desirable. (London Health Sciences Centre, 2005) Therefore, resistance to the concept of tissue donation is not expected (Nather, 2000).

In fact, many Hindus believe that tissue donation is in keeping with Hindu beliefs. *Daan* refers to selfless giving in donation in Sanskrit. As such, to be able to donate one's tissues and organs for the benefit of another is deemed by many Hindus as a good practice.

> *Of all the things that it is possible to donate, to donate your own body is infinitely more worthwhile.*
>
> The Manusmruti, the most important and earliest metrical work of the Dharmaśāstra textual tradition of Hinduism

LEGAL CONSIDERATIONS

There is no universal law governing tissue procurement and tissue transplantation for the various countries in the Asia-Pacific region. If regulatory laws are present in some of these countries, they are based on similar human transplantation acts practiced in Europe and the United States. These acts cover a wide range of issues, which include the definition of brain death, the definition of tissues and organs, the issue of consent for organ donation (either by the donor or next-of-kin), and the prohibition of trade in human organs.

Two different legal frameworks are seen to be operating in the Asia-Pacific region: the "opting-in" system based on informed consent, and the "opting-out" system based on presumed consent. For example, Malaysia follows the opting-in system for kidneys, liver, heart, lungs, cornea, skin, heart valves, and bones. Singapore follows the "opting-out" system under the 1987 Human Organ Transplant Act for kidneys, cornea, liver, and heart only. However, Singapore follows the opt-in system for other organs such as the skin and lungs under the Medical (Therapy, Education and Research) (MTERA) Act. It should be noted that in almost all countries, these laws are specifically designed for organ transplantation. Tissue procurement can be carried out only by following such laws for organs. The systems operating in each of the countries will now be described.

Singapore

It is stated that tissue procurement follows the MTERA Act of 1973, whereby "any person of sound mind and eighteen years of age or above may give all or any part of his body for education … transplantation. ... The gift takes effect upon death" (Nather, 2000). The MTERA Act is an opt-in system for people aged 18 and above to allow their organs and tissues to be used in transplantation for the benefit of a recipient, in education or in research.

In 1987, the Human Organ Transplant Act (HOTA) was passed. It is based on an opt-out system which presumes consent. It "makes provision for the removal of kidneys from the bodies of persons who are citizens or permanent residents, who have died from accidents, for transplantation purposes only. Muslims and persons over 60 years old are exempted from provisions of this Act." The seven criteria for brain death are also described in this Act. An amendment to the Act in 2004 extended the list of tissues procured to include the heart, liver, and cornea, and extended the donor pool to non-accidental causes of death, among other changes. With effect from 2008, Muslims were also included in the HOTA Act.

Under the HOTA Act, people who have opted out of the system are also placed on a lower position on the waiting list should they come to need an organ in the future. Singapore also follows the "hard" opt-out system which means that the family member cannot override the patient's decision to donate his organs upon the patient's death. In order to ensure that Singaporeans can make an informed decision as to whether they would like to opt out of HOTA, the Ministry of Health currently carries out public education programs. These include a letter sent to citizens when they turn 21, roadshows, and articles in the newspaper twice a year.

Australia

Australia's current organ donation system is based on expressed consent or an "opt-in" system. Potential donors must register their intent to donate organs under the Australian Organ Donation Registry. As such, Australia often faces a shortage of organ and tissue donors. To combat this problem, the Federal Government set up the Australian Organ and Tissue Authority under the Australian Organ and Tissue Donation and Transplantation Authority Act 2008. The objective of the Authority is to implement nationwide programs to help increase the tissue and organ donation rate across the country. The Authority hopes to increase organ and tissue donation rates through public education on organ donation as well as increasing public confidence in the organ transplantation process.

Under Australian law, organs that can be transplanted include the liver, heart, lungs, kidneys, intestines, and pancreas. Tissues that can be transplanted include the bone, tendon, ligaments, skin, heart valves, and other heart tissues, as well as various parts of the eye such as the cornea and sclera. Organ and tissue donation can be performed when the donor is alive (in the case of liver and kidneys) and when the donor is dead (for all other organs and tissues as mentioned above). There are two different definitions of death: brain death or circulatory death. The patient's age, medical history, and nature of death will be considered as to whether the patient's organs and tissues are eligible for donation. The patient's family as well as any prior instructions the patient may have given with regard to organ donation will be taken into account before organs are removed from the patient.

The donation of human tissue is regulated by the legislation in each of the eight states and territories under substantially uniform acts (known as the Human Tissue Act in some states, and as the Transplantation and Anatomy Act in others), which were passed in the late 1970s and early 1980s. Most provisions require consent from the donors or from the families of brain-dead heart-beating donors, with the exception of tissues removed at autopsies that can be used for transplant, therapeutic, educational, and research purposes without further reference to the next-of-kin. Before organs are transplanted, the patient's family will be consulted. In some cases where families have already discussed the issue of organ donation before, families will already know the loved one's decision. In other cases where families have not discussed the issue of organ donation, families will need to make a decision whether or not to donate their loved one's organs. Nonetheless, the tissue banking sector in Australia has, for the most part, sought to include consultation with next-of-kin in their protocol for practical and ethical purposes (Ireland and McKelvie, 2003).

India

Under Indian laws, tissue and organ donation follow an "opt-in" system, where people who wish to donate their organs have to record their wishes on a registrar and ensure that their decision is made known to their family. The procurement of tissues for transplantation is governed by the Transplantation of Human Organs Act, which was enacted in 1994. Under the Act, brain death came to be considered a form of death in which organ transplantation was acceptable. Also, the Act made organ trading a punishable offence. However, as healthcare comes under the purview of state governments, this Act is applicable only when a state adopts it. As of 2011, many Indian states have adopted the law and further amendment of the law has taken place. This has helped to slowly increase the number of organ donations from patients upon brain death (see Table 6.1).

A human organ as defined by this Act is any part of the human body consisting of a structured arrangement of tissues, which cannot be replicated by the body if wholly removed. There is no specific mention of particular organs. Under the Act, therapeutic purposes are defined as the systemic treatment of any disease or the measures to improve health according to any particular method or modality. The Act recognizes brain stem death, which needs to be certified by four qualified medical practitioners approved by the state. The removal of organs is subject to prior written consent from the deceased or from the person who is lawfully in possession of the dead body, with the exception of the removal of organs from bodies sent for post-mortem examination. In the latter case, the Act

Table 6.1. Deceased organ donation in India — 2012.

State	No. of deceased donors	Total no. of organs retrieved	Organ donation rate per million population
Tamilnadu	83	252	1.15
Maharashtra	29	68	0.26
Gujarat	18	46	0.30
Karnataka	17	46	0.28
Andhra Pradesh	13	37	0.15
Kerala	12	26	0.36
Delhi-NCR	12	31	0.29
Punjab	12	24	0.43
Total	**196**	**530**	**0.16**

authorizes the removal of organs for therapeutic purposes, provided there were no known objections from the deceased person.

While regulations specific to tissue banking have yet to be developed, the Transplantation of Human Organ Rules was issued by the Government of India in 1995 to combat the illegal trading of human organs. It has since been adopted by the state of Maharashtra. However, while organ trading is illegal in India, there have still been reported cases of illegal organ trafficking in India (Gajiwala 2003).

Malaysia

The transplantation of cadaveric tissues in Malaysia is governed by The Human Tissues Act 1974. The Act enables the removal of tissues from cadavers for therapeutic, medical education, and research purposes under two conditions: at the express request of the donor, which may be given at any time either in writing or orally stated during the deceased's last illness in the presence of two witnesses; or in the absence of objection from the deceased and with the consent of the next-of-kin (Kassim, 2005).

A problem faced in Malaysia is the lack of cadaveric donors for the large demand of organs required. This is further complicated by the ambiguity within The Human Tissues Act of 1974. In this Act, the word "tissue" is not defined. Likewise, "the person lawfully in possession of the body" is not defined, nor is there an articulation of a hierarchy of relatives deemed to be the next-of-kin. More significantly, the current Act does not provide an exact definition of death. Presently, the Act only requires two fully registered medical practitioners to confirm (upon personal examination of the body) that life is extinct. There is no inclusion of brain death in this Act as a method of determining death, although brain-dead donors are a source of organs in cadaveric organ transplantation. With regard to *fatwas* specific for tissue donation, the first *fatwa* on bone, skin, and amnion was introduced by the Malaysian Islamic Center on September 4, 1995 (Nather, 2000).

In order to raise public awareness on organ and tissue donation to secure more organs, the National Transplant Research Center was set up by the Ministry of Health, in cooperation with the Medical Development Division. Its main aims are to spread correct information about organ donation, getting more people to pledge and register as organ donations, and collaborating with various non-governmental organizations, religious bodies, and student associations to spread information on organ and tissue donation to the public.

Sri Lanka

Sri Lanka follows The Human Tissue Transplantation Act No. 48 of 1987, which requires consent from the donor or next-of-kin. Under this Act, persons aged 21 and above are allowed to give consent for any part of his body to be used for transplantation or research and educational purposes upon death. This consent can be given by the donor either orally or through written forms in the presence of two or more competent witnesses.

Philippines

Tissue donation in the Philippines follows The Republic Act 7170, 1991, which authorizes the donation of all or part of the human body after death for specified purposes with a person's consent (Nather, 2000) A person is able to make a decision as to whether he would like to donate his organs upon death when he has reached 18 years of age and is of sound mind. In cases where the individual has not specified whether he would like his organs donated upon death, Filipino law allows for the family to make the decision on the individual's behalf when the individual is no longer able to do so.

Vietnam

Tissue procurement is provided for by the Civil Code, Article 32, Chapter 2, where consent is needed from the donor or next-of-kin; and by The People's Health Protection Code, Chapter 4, which provides for tissue transplantation (Nather, 2000).

Indonesia

In the Asia-Pacific region, Indonesia is unique in that its legislation for tissue procurement is incomplete. The Indonesian 1992 Health Regulation provides for the procurement of tissues from living donors, but not from deceased donors. A *fatwa* for bone, skin, and amnion was introduced by the Religious Council on June 29, 1997, permitting tissue procurement from cadaveric donors (Nather, 2000).

Japan

In Japan, a new law concerning human organ transplants was passed on October 17, 1997 (Ota, 1999). According to the Law, organs can be removed only if the donor has expressed his or her intention with respect to the definition of death and organ donation in a written document beforehand, and if the family has already signed the donor card and also agreed with organ removal (the family has the authority to veto an individual's organ donation by refusing to sign the donor card). Brain death is defined under this Law as "an irreversible cessation of all functions of the entire brain including brain stem." Under the Law, individuals are able to choose a definition of death, either brain death or traditional cardiac death, according to their own personal views on human death.

As of December 2005, only 33 brain-dead cases have been used for organ transplantation (Bagheri, 2005). The first of these cases was performed on February 28, 1999, when a liver, a heart, and two kidneys were legally procured from a brain-dead donor in Kochi and air-flown for transplantation to four recipients in four other Japanese cities (Reuters/AFP, 1999). It was hoped that this landmark event would change the legal environment to favor tissue transplantation in Japan (Nather, 2000). Unfortunately, this has not happened.

Japan has a significantly lower rate of organ donation as compared to Western countries. In fact, it has the lowest organ donation rate among all developed countries. This is due to various cultural reasons involving life, ethic, and death, as well as a relative distrust in Western medicine. The predominant religion in Japan — Shinto — regards death as impure, and this attitude and belief have led to various prejudices against tissue donation in Japanese culture and religion.

Korea

The Organ Transplantation Act was passed in Korea in 2000. Under this Act, brain death is defined as the irreversible cessation of the whole brain function, and has to be diagnosed by two specialist doctors and the patients' physician as well as approved by a brain death determination committee. Donor consent is required for organ removal, but the family has strong veto power over organ transplantation. The Korean Network for Organ Sharing is a centralized authority for organ procurement. Since the Act, the number of brain death diagnoses and donations has decreased. Before the Act was enforced, 162 cases were diagnosed as brain-dead; as of 2003, only 43 cases were diagnosed (Bagheri, 2005).

To cope with the decrease in the number of brain-dead patients and the increasing demand for organs, the Organ Transplantation Act was revised by the government in August 2002, with the introduction of an incentive system. Under this system, if a transplantation hospital formed a Committee for Brain Death Evaluation and a Hospital Organ Procurement Organization, it could receive a kidney from a brain-dead donor as an incentive to foster organ procurement regardless of the Korean Network of Organ Sharing wait list. The government also launched a pilot brain-dead donor registry program to strengthen Hospital Organ Procurement Organization activity.

Following a successful operation on 6-year-old Jo Eun-so in 2012, the government also legalized the donation of the stomach, duodenum, colon, and spleen. Previously, only nine organs were legalised for transplantation in Korea. These organs include the kidneys, liver, pancreas, small intestines, and bone marrow. The Korean Organ Donor Program was established in 1991 to promote greater public awareness on organ and tissue donation, as well as to secure more organ donors.

Wales

Like much of the United Kingdom, organ donation in Wales used to follow an opt-in system. However, due to the need for more organs, Wales recently converted to a soft opt-out system in 2013, which will take effect from December 1, 2015.

Under the soft opt-out system, all people aged 18 and above living in Wales will be presumed to have given consent to have their organs donated to suitable recipients upon their death. As Wales follows a soft opt-out system, the individual need not explicitly choose whether he would want to opt in or opt out while he is living. In such instances, the family's opinion will also have to be consulted before the deceased's organs can be donated.

In order to prepare the Welsh public for the changes in organ donation in Wales, the government has launched a nationwide program entitled "It's About Time". In its initial phase, the program will focus on digital platforms before being rolled on to other forms of media. The campaign will provide the Welsh with information about the changes implemented with regard to organ donation as well as encourage people to discuss their wishes with family.

Bangladesh

The Tissue Donation and Transplantation Act was passed in April 1999, permitting donation from live and cadaveric donors (Nather, 2000). The Act also prohibits the trading of organs.

Countries in the Asia-Pacific Region with No Laws on Tissue Donation

In contrast, several countries in the Asia-Pacific region do not have any law concerning tissue procurement and transplantation. Such countries include Thailand, China, and Myanmar. In China, although there is no Human Transplantation Act, transplantation is nevertheless practiced in accordance with the principle of written consent for donation prior to the patient's death or from the next-of-kin. In Myanmar, apart from the Eye Donation law, no law for tissue donation exists (Shroff, 2002). In Thailand, no law exists. However, the Medical Council is responsible for regulating human organ transplantation. They decide the criteria from time to time, and determine whether any punitive action should be taken against the doctors.

CULTURAL ISSUES PERTAINING TO ORGAN DONATION AND TRANSPLANTATION

A common principle followed by all religions of the world is that saving lives overrides all objections. What holds people back from organ donations and transplantation are cultural reservations such as ignorance of the process of organ donation, fear of mutilation, and fear that organs will be sold or used only for the rich. Hence, certain cultural issues have to be resolved before organ donation and transplantation can progress in any country.

Muslims culturally accept that God created them whole and they prefer to return to Him whole. However, Islam does not explicitly forbid donation. However, the cultural practice among many Muslims is to bury amputated limbs, and even foreskin from circumcision and amnion from delivery. There is thus a big shortage of bone donors in Muslim countries. Likewise, Christianity promotes organ and tissue donation. However, there is a shortage of donors in Christian-dominant countries like the Philippines, due to the cultural attitudes of the people who also prefer to return to God whole.

To overcome cultural barriers, there is a need for more public awareness programs on the need for and benefits of tissue donation and tissue transplantation in the community. As mentioned previously, education programs in different counties from Korea to Wales are increasingly taking root, with the aim of better informing citizens about the need for and benefits of organ donation. For example, it has been found that in cases where the wishes of the deceased are not known, only 50% of their relatives would consent to tissue donation. Yet, it is shown that through encouraging people to talk and discuss organ donation and tissue transplants with their relatives, the number of people consenting to organ donations and tissue transplants could increase by 93–94% (Shroff, 2002). Hence, there is a need to address the cultural and religious issues that may hinder tissue donation. In addition, there is also a need for professional awareness among doctors to encourage more surgeons to carry out more tissue transplantations.

REFERENCES

American Association of Tissue Banks (2007) About AATB, available at http://www.aatb.org/.

Ancoq AC Phillipine law on donations of human organs, *J Int Bioethique* **3**, 169–171.

Bagheri A (2005) Organ transplantation laws in Asian countries: a comparative study, *Transplant Proc* **37**, 4159–4162.

El-Shahat YIM (1999) Islamic viewpoint of organ transplantation, *Transplant Proc* **31**, 3271–3274.

European Association of Tissue Banks (2003) Common Standards for Tissues and Cells Banking. EATB, available at http://www.eatb.de/.

Gajiwala AL (2003) Setting up a tissue bank in India: The Tata Memorial Hospital experience, *Cell Tissue Bank* **4**, 193–201.

Hachiya Y, Sakai T, Narita Y, Izawa H, Yoshizawa K (1999) Status of bone banks in Japan, *Transplant Proc* **31**, 2032–2035.

Human Tissue Act (2006) HTA, available at http://www.hta.gov.uk/about_hta.cfm.

Ireland L, McKelvie H (2003) Tissue banking in Australia, *Cell Tissue Bank* **4**, 151–156.

Kassim PN (2005) Organ transplantation in Malaysia: A need for a comprehensive legal regime, *Med Law* **24**, 173–189.

Laws of Malaysia (1974) Act 130: Human Tissues Act, 1974, Ketua Pengarah Percetakan.

Nather A (2000) Tissue banking in Asia Pacific region: Ethical, legal, religious, cultural and other regulatory aspects, *J ASEAN Orthop Assoc* **13**, 60–63.

Ota K (1999) Proceedings of the 27th World Congress of the Transplantation Society, *Transplant Proc* **31**(1), 205–209.

Reuters/AFP (1999) Media frenzy shocks heart-donors family: Breaking of taboo. *The Straits Times,* March.

Shroff S (2002) Cadaver-Organ-Donation, available at http://www.medindia.net/slides/ppt/Cadaver-Organ-Donation.ppt.

Transplantation of Human Tissues Act Arrangement of Sections (2013) Retrieved May 14, 2014, from http://srilankalaw.lk/revised-statutes/volume-viii/1318-transplantation-of-human-tissues-act.html.

7

CHALLENGES FOR TISSUE BANKING IN THE ASIA-PACIFIC REGION AND RECENT PROGRESS IN MALAYSIA

Norimah Yusof

INTRODUCTION

In most parts of the world, tissue banking had a humble beginning, and seemed initially an in-house bone-collecting effort for hospital uses. Phillips (2013) described tissue banking as a "cottage industry" where individual hospitals rescued pieces of bone and skin from surgical procedures and, after minimum treatment, used the tissues in patients. Tissue banks needed to have better documentation since increasing demands for tissue warranted proper and traceable records. As for other medical activities, the tissue bank needs to have all levels of its activities checked, including the tissue quality throughout the processing line. It is therefore pertinent for tissue banks to adopt a quality system in their management. Setting up the quality system needs a tremendous effort, requiring a committed organisational structure, trained personnel, validated procedures, well-equipped facilities and continuous financial support. Such requirements are essential to achieve technical recognition for the safety of tissues produced. Ultimately, the bank must comply with local regulatory requirements where needs for certifications from local authorities become a compulsory if not on a voluntarily basis. As tissue banks get more established and come under the purview of local governmental agencies and international bodies like the World Health Organization (WHO), more formal legislation is required.

National legislation is often consistent with international regulations as a result of international harmonisation, a significant contribution which has come from WHO. Thus, tissue banks have to organise all internal processes and implement them in such a way that they conform to the national laws and regulations. The banks must be able in the same manner to adopt any standards being developed by any international bodies that are recognised by the local authority. The tissue banking community must be alert to new findings and advances developed in international scientific circles. Tissue banking remains a regional or national activity in the Asia-Pacific region but has gained a great deal from the coordination efforts of the International Atomic Energy Agency (IAEA).

REGIONAL PROGRAMMES

Tissue banking in the Asia-Pacific region started in the early 1980s with the first bank established by Dr U Pe Khin in Rangoon, Burma, followed by Bangkok Biomaterial Centre by Prof Yongyudh Vajaradul in Bangkok, Thailand, in 1984 (Phillips, 2013; Nather and Zheng, 2010). Despite this early start, as compared to other regions likes Latin America, tissue bankers in the region had to deal with ethical, religious, and cultural issues that hindered the development of tissue banking; these issues had to be carefully handled in promoting any transplantation programme. The Technical Assistance Programme on Tissue Banking initiated by IAEA in 1984 and the Regional Cooperative Agreement (RCA) Project IAEA/RAS 7/008 on Radiation Sterilization of Tissue Grafts in the following years (1985 to 2004) succeeded in introducing the tissue banking concept among medical specialists, nuclear researchers and lawmakers in the Asia-Pacific region. Low-cost tissue processing together with efficient radiation sterilisation technology introduced by IAEA experts encouraged the development of many tissue banks in Member States. Close collaboration among medical professionals and research scientists has led to the establishment of many banks. By 2000, almost every country in the region — China, India, Indonesia, Bangladesh, the Philippines,

Hong Kong, Korea, Malaysia, Myanmar, Pakistan, Singapore, Sri Lanka, Thailand and Vietnam — had at least one operating tissue bank that could supply tissue grafts, and thus reduce the importation of tissues which is costly and poses a risk of disease transmission when imported from abroad.

This remarkable achievement by IAEA in establishing tissue banks in developing countries has been beneficial to those patients requiring tissue transplantation for better healthcare, while creating awareness of tissue donation for the society at large. The application of radiation in reducing the risk of disease transmission opened up another opportunity for nuclear technology to be exploited for human kind. In 2000, as proposed by Professor Glyn O. Phillips, a survey was carried out among the project counterparts to quantify the success. More than 220,000 tissue grafts had been produced and utilised clinically in 16 countries of the Asia-Pacific region (Morales and Phillips, 2004). The value of the grafts was estimated at US$51,768,553, based on supply cost by many banks in the United States and Europe, and this was higher than the project cost spent by the IAEA in the same period of time, i.e., US$6,313,335 including US$2,036,034 specifically allocated for training programmes. Tissue grafts produced under the IAEA programme from 1998 to 2000 by the 16 countries were mainly freeze-dried cancellous bones (69,195), large frozen bone allografts (8,588), skin and amnion (96,645) and other tissues including tendon, ligament, fascia lata, pig skin, duramater, demineralised bone and cow pericardium (50,278). The benefits of the programmes are listed in Table 7.1.

REGULATORY SYSTEM

Tissue banks should have internal set-ups with an organised structure to operate and implement their function, which respects all national laws and regulations or any international laws adopted by the healthcare authorities in the field of tissue banking. Tissue banks must also identify appropriate and recognised standards to follow. As shown in Figure 7.1, a tissue bank must be aware of laws and regulations that exist and that must be complied with, as well as standards and guidelines that could assist in producing safe and quality tissues.

Table 7.1. Benefits from the IAEA programme on radiation sterilisation of tissue grafts in the Asia-Pacific region.

- Improvement in healthcare in many Member States;
- Avoidance of any complications due to infection and avoidance of complications at secondary site for autograft;
- Overcoming insufficient grafts when autograft is used, especially in cases involving children and babies;
- Skin and amnion acting as wound coverage and burn dressing that can enhance healing, while giving protection from infection and minimising water loss;
- Opening up wide opportunities for young surgeons in developing countries to learn new approaches in treating patients with tissue transplantation;
- Exchanging of knowledge in surgical treatment between surgeons through close collaboration;
- Through conferences and meetings, discussion of various common issues in relation to tissue banking, including legal and technical aspects.

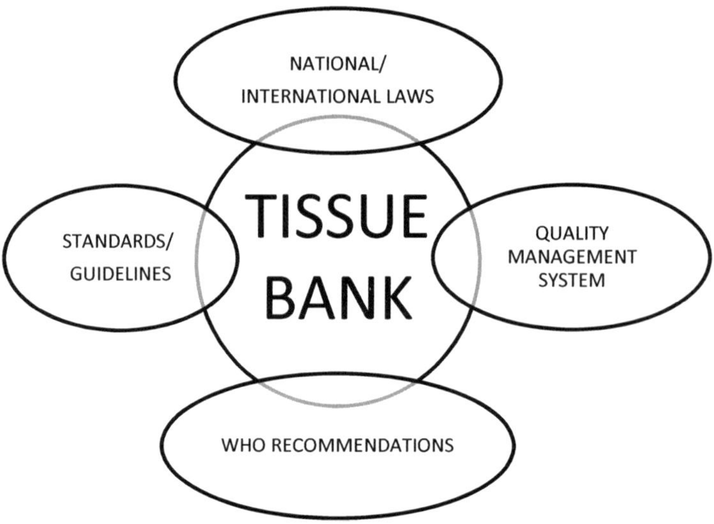

Figure 7.1. Linkages of a tissue bank with regulatory and quality systems and standards.

Procurement and Transplantation Laws/Acts

As reported by Nather *et al.* in 2010, not all of the countries in the Asia-Pacific region have laws directly related to tissue banking. In Australia, Bangladesh, Hong Kong, Indian, Malaysia, Sri Lanka and Vietnam, the requirements of the Human Tissue Acts are similar to Human

Transplantation Acts in Europe and in the United States. Singapore and the Philippines have their respective Acts which allow the donation of any part of the body including both organs and tissues.

The legal status governing tissue procurement and tissue transplantation for various countries in the region is summarised in Table 7.2. In countries where the laws are specifically designed for organ

Table 7.2. Legal status governing tissue banking in the Asia-Pacific region.

Country	Legal status
Australia	Donation of human tissues is regulated by the legislation in each state and territory under substantially uniform Acts known as the Human Tissue Act in some states and as the Transplantation and Anatomy Act in others, which were passed in the late 1970s and early 1980s. Most of the provisions require consent from the donors or from the families of brain-death, heart-beating donors with the exception of tissues removed at autopsies that can be used for transplant, therapeutic, educational and research purposes without further reference to the next-of-kin. However, most tissue banks attempt to include consultation with the next-of-kin in their protocol for practical and ethical purposes.
Bangladesh	Donation from live and cadaveric donors is permitted under the Human Organ/Tissue Donation and Transplantation Act which was passed in 1999 requiring consent from the donor or next-of-kin. The Safe Blood Transfusion Act, 2002 makes serological screening mandatory for tissue/organ donation.
China	Although China has no Human Transplantation Act, written consent is obtained from donors before procurement.
Hong Kong	Donation is regulated by the Human Organ Transplantation Ordinance 1997, an "opting-in" law requires consent from the donor or next-of-kin.
India	Donations previously followed the Human Anatomy Act 1949 which covers the recovery of tissues from unclaimed or donated bodies for therapeutic purposes, medical education or research. Under the Transplantation of Human Organs Act enacted in 1994 and amended in 2009 to include tissue banks, written consent is required from the donor of not less than 18 years of age with individual consent before death in writing and in the presence of two witnesses, of whom at least one must be a close relative of the person, or the person lawfully in possession of the dead body. This Act is applicable only when a State adopts it since healthcare comes under the purview of State governments. The human organ as defined by the Act is any part of the human body consisting of a structured arrangement of tissues which cannot be replicated by the body if wholly removed. The Act also recognises brain stem death. Transplantation of Human Organ Rules were issued in 1995 to curb the illegal trading of human organs.

(*Continued*)

Table 7.2. (*Continued*)

Country	Legal status
Indonesia	The Health Regulation 1992 provides an "opting-in" system that allows retrieval of tissues from living donors only. A *fatwa* on bone, skin and amnion enacted in 1997 by the Religious Council (Majelis Ulama Indonesia, MUI) permits tissue procurement from live and deceased donors. The *fatwa* is not legally binding.
Japan	Cornea is the only tissue legally procured under the Law for Transplantation of Kidneys and Corneas 1979. Legislation on brain death came into force in 1997 under the Organ Transplantation Law. Legislation of Advanced Medical Treatment 2004 allows approved institutions/bone banks to charge the expenses of authorised treatment. Cryo-preserved allogenic tissues from cadaveric donors are approved as advanced medical treatment in 2007.
South Korea	The Organ Transplantation Act 2000, reviewed in 2002, covers organs and tissue (only cornea). Brain death is defined and legally recognised. Brain death must be diagnosed by two medical specialists and the patient's physician, and must be approved by a brain-death determination committee. Donor consent is required, but the family has a strong voice to not allow organ donation if they disagree.
Malaysia	The Human Tissues Act 1974 allows donation as an "opting-in" system which was initially given to kidneys and corneas. The Act enables the removal of tissues from cadavers for therapeutic, medical education and research purposes. A *fatwa* on bone, skin and amnion was issued in 1995 by The Malaysian Islamic Centre allowing Muslims to donate. The National Organ, Tissue and Cell Transplantation Policy was developed in 2009.
Myanmar	There is no tissue donation law apart from the Body Organ Donation Law issued in 2004 for organ transplantation, research and education. The Law only mentions eye donation, thus the Law needs revision. A person who has attained the age of 18 may donate his own body organ. The next-of-kin may donate the deceased's body organ if there is no objection in the family.
Pakistan	Transplantation of Human Organs and Tissues Act 2009, enacted in 2010, is to provide for removal, storage and transplantation of human organs and tissues for therapeutic purposes (The Gazette of Pakistan, 2010). The Act includes donation from the living and cadavers. A living donor who is not less than 18 years of age may during his lifetime voluntarily donate any organ or tissue.Government efforts, supported by professional associations, civil society organisations and the media, along with WHO technical assistance, have led to the development of this legislation and to curbing organ trade in conformity with international guidelines.

(*Continued*)

Table 7.2. (*Continued*)

Country	Legal status
Philippines	Donation is allowed by the Republic Act 7170 (Organ Donation Act) in 1991 as an "opting-in" law whereby donation of all or part of a human body after death including brain death for specified purposes must be authorised by the donor or next-of-kin. An international sharing of human organs or tissues is allowed and shall be made only through exchange programmes duly approved by the Department of Health, provided that the foreign organ or tissue bank grants reciprocal rights to their Philippine counterparts to draw organs or tissues at any time.
Singapore	Donation is allowed under the Medical (Therapy, Education and Research) Act of 1972, whereby any person of sound mind and 18 years of age and above may give all or any part of his or her body for education or transplantation. As an "opting-in" system, written consent from donor or next-of-kin was required. Later, Singapore followed the "opting-out" system under the 1987 Human Organ Transplant Act (HOTA) for kidneys only and then extended it to include the heart, liver and corneas after amendment in 2004. The Act describes the seven criteria for brain death.
Sri Lanka	Donation is allowed under the 1987 Transplantation of Human Tissue Act No. 48. As an "opting-in" law, written consent from the donor or next-of-kin is required. Any person above the age of 21 years may consent to the donation, to take effect upon his death, of his body or any part or any tissue to be used for the advancement of medical and dental education, for the purposes of research, for the advancement of medical science or therapy or for use on any living person
Thailand	No tissue donation law is issued and there is no law to directly govern transplant procedures.
Vietnam	Tissue banks abide by two laws, namely Article 32 of the Civil Code, Chapter 2 on tissue donation where consent is needed from the donor or next-of-kin; and the People's Health Protection Code, Chapter 4 on tissue transplantation.

Sources: Bagheri, 2005; Bile *et al.*, 2010; Gajiwala, 2003; Hilmy *et al.*, 2009; Joo, 2013; Kassim, 2005; Ireland and McKelvie, 2003; Nather *et al.*, 2010; Ohkawara *et al.*, 2010; Schoff, 2005; Zahid *et al.*, 2010.

transplantation, tissue procurement can be carried out only by following the laws specified for organs. The Acts address the definition of brain death, the definition of organs and tissues, the issue of consent for donation and the prohibition of trade in human organs and tissues. Most countries have some laws for tissue procurement but not all have brain death legislation in place yet (Shroff, 2005). This might be the reason why cadaver donors per million population per year for Asia was the lowest (1.1) as compared to the United States (20.7), Europe (15.9) and South America (2.6) (Shroff, 2005).

Each country has its own legal status for tissue transplantation but with different levels of enforcement. Most of the countries in the Asia Pacific region have an "opting-in law" based on informed consent except Singapore that has switched to an "opting-out" system based on presumed consent. Under the Singapore Human Organ Transplant Act (HOTA), Act 15 of 1987 amended in 2004 (Nather *et al.*, 2010; Bagheri, 2005), the designated medical doctor can authorise in writing the removal of any organ from the body of a person who has died in hospital for the purpose of transplantation, except for the following persons:

- who have during their lifetime registered an objection to the removal of the organ after his death (presumed consent);
- who is neither a citizen nor a permanent resident of Singapore;
- who is below 21 years of age unless the parent or guardian has consented to such removal; and
- who is above 60 years of age; and who is a Muslim.

Tissue banks should not "buy" or "sell" tissues after processing. Charges, if imposed on patients, should only be to cover the cost of processing, storage and transportation for despatch. Therefore, most of the tissue banks in the region are established in public hospitals, university hospitals or nuclear centres with financial support from the government. However, in the late 2000s, a few of the banks have been privatised such as Shanxi Provincial Tissue Bank in China which is commercially run by OsteoRad Biomaterial. In addition, several private tissue banks have been established such as by Australian Biotechnologies in Sydney, and Bioland and Hans Biomed in South Korea.

The legal status related to importation of human tissues is another aspect of legislation that is still not widely available in many countries. Safety control systems and legal penalties are future issues to be looked into. Buying and selling of human tissue grafts are strictly prohibited; however, exporting or importing of the grafts for transplantation has been ongoing, presumably with permission from the regulatory authority.

Laws and Regulations for Tissue Banks

Only a few countries have established laws and regulations specific to tissue banking. In general, it seems laws specifically on tissue are not widely available as compared to organs. Although a law might exist, it is primarily related to organ donation. In some countries, human tissue is excluded, and tissue donation is left to its own self-regulation. Tissue donation in such instances is an act of humanity and solely on a voluntary basis. There is no universal law governing tissue procurement and tissue transplantation. WHO has set up guidelines to guide any country considering developing their own laws and regulations (WHO, 2010). Some countries in the Asia-Pacific region have formulated regulatory laws based on those practised in Europe and the United States.

Tissue banks in **Australia** have to comply with the Australian Code of Good Manufacturing Practice for Therapeutic Goods — Human Tissues, first issued in 1995; its latest revision was released in 2014. This Code requires tissue banks to establish quality systems according to ISO 9000 (Nather *et al.*, 2010). Tissue banks must apply for yearly licensing from the Therapeutic Goods Administration (TGA) whereby annual audits have to be performed. High standards for tissue banks have been established with which tissue banks must comply. Australia is one of the few countries with defined strict guidelines in place to assess donor eligibility for patients who are considered brain-dead, but are still on life support (Zheng *et al.*, 2006). The issue related to cadaveric donors raises a number of challenging legal and ethical problems whereby the determination of death and organ retrieval needs a quality assurance system for cadaveric donation with input from medical teams, regulating agencies and tissue banking associations (Zheng *et al.*, 2006).

In **China**, organ transplant practice has developed greatly in the last two decades even though no act has been introduced. The Regulations on Human Organ Transplant 2007 was introduced to replace the previous Interim Provisions on Administration of Clinical Application of Human Organ Transplant Technology 2006 (Ding, 2008). The Regulations have some impact upon three problems existing in the current organ transplant practice, that is, organ procurement from executed prisoners, organ trade and organ tourism. The deficiencies of the 2007 Regulations are supposed to be improved in the next legal reform (Ding, 2008).

In **Japan**, although organ procurement has been regulated by The Organ Transplantation Law (brain-dead donors since 1997, donors after cardiac death since 1979), there is no law or governmental procurement network except for corneas (Ohkawara *et al.*, 2010). Since the late 1980s, some university hospitals have developed their banks. In 2001, the Japanese Society of Tissue Transplantation (JSTT) and Japan Tissue Transplant Network (JTTN) established guidelines for tissue procurement with the aim to coordinate tissue harvesting. Nevertheless, it remains dramatically behind Western countries.

In **South Korea**, tissue banks should obtain permission from the Korean Food and Drug Authority (KFDA) to operate legally. The banks should be well equipped with standard equipment; the facility must have adequate manpower that are trained in tissue banking; and it must have a quality system in place that satisfies all the requirements set by the KFDA. A new law introduced in 2004 and partly amended in 2005 facilitated investigation on safety of tissues. However, there are still problems caused by imported grafts. Since 2005, a few banks have been charged for violating standards and several hospitals have faced penalties for violations and have even been suspended. Under the Organ Transplantation Act (2000), the Korean Network for Organ Sharing (KONOS) was established with centralised authority for organ procurement as well as for approval of donors and recipients (Lee and Kim, 2009). However, the number of brain-dead donors decreased sharply, which led the government to revise the Organ Transplantation Act in August 2002; an incentive system was introduced and a pilot brain-dead donor registry programme was launched to strengthen Hospital Organ Procurement Organization activity (Lee and Kim, 2009).

In **India**, the Transplantation of Human Organs Act (amended in 2009) has been renamed the Transplantation of Human Organs and Tissues Act to provide the regulation for removal, storage and transplantation of human tissues for therapeutic purposes and for the prevention of commercialised dealings in human tissues. Tissue banks or the institutions responsible for tissue banking have to be registered under the Ministry of Health and Family Welfare (MOHFW). The word "tissue" has been defined as a group of cells (except blood) performing a particular function in the human body. It is mandatory for a transplant coordinator to be engaged in a hospital performing organ and tissue transplantation. More amendments were made in 2011, the following among others:

- The central government may establish a National Human Organs and Tissues Removal and Storage Network at one or more places, and Regional Network.
- The central government shall maintain a national registry of the donors and recipients of human organs and tissues.
- Parties that deal with illegal dealings of human tissues will be punished.
- A tissue bank must be registered.

No tissue bank shall commence any activities relating to tissue banking unless it is duly registered under this Act. Tissue banks have to be registered by the appropriate authority that will determine if a bank is in the position to provide such specialised services and facilities; if it possesses skilled manpower and adequate equipment; and if it can maintain standards. The government will introduce a regulation that describes the function of tissue banks. In 2014, the central government released the Transplantation of Human Organs and Tissues Rules which shall come into force on the date of their publication in the Official Gazette (The Gazette of India, 2014).

The legal status of tissue banking in **Indonesia** was made clear in 2009 after the government released UU No. 36 which governs the procurement, processing and transplantation of biological tissues for implantation (Republik Indonesia, 2009). Recovery and healing from disease are permitted through organ transplantation, tissues, implants, medical devices, plastic/reconstructive surgery and application of stem cells. The transplantation is allowed only for humanitarian, not commercial, purposes. Trade in organs and tissues is absolutely forbidden. Organ and tissue transplantation must be conducted by medical officers who have the ability and authority to do it and must be carried out in medical facilities. For procurement of organs and tissues, attention must be given to the donors' health and consent must be obtained from the donors and/or their next-of-kin. Regulations on anatomical cadaveric surgery and cadaveric tissue transplantation are now being prepared.

In **Myanmar**, under the Body Organ Donation Law 2004 (The Union of Myanmar, 2004), whoever sells or buys the body organ for consideration shall, on conviction, be punished with imprisonment for a term which may extend to three years, and may also be liable to a fine. Regulations for tissue banking are not yet available.

In **Thailand**, despite not having any specific law, the Medical Council is responsible in regulating human organ transplantation and decides the criteria from time to time. The Medical Council determines whether any punitive action should be taken against errant doctors (Schoff, 2005).

CHALLENGES FOR TISSUE BANKING IN MALAYSIA

The challenges and progress of legislation for tissue banking in Malaysia will now be considered. The first tissue bank in Malaysia was established in the early 1990s at Universiti Sains Malaysia (USM) Medical Centre by the surgeon Dr Hasim Mohamad supported by a group of researchers from Malaysia Nuclear Agency led by Dr Norimah Yusof. The USM Tissue Bank was inaugurated in 1994 and has been expanding ever since from a small four-room bank to a certified bank equipped with cleaned rooms set up in 2013 under the leadership of Dr Suzina Sheikh Abdul Hamid. In the following years, two bone banks were established at Hospital Kuala Lumpur and University Malaya Medical Centre in 1998 and 1999, respectively.

Laws for Tissue Banking in Malaysia

Transplantation of cadaveric tissues in Malaysia is governed by the Human Tissues Act 1974 (Laws of Malaysia, 1974) as described in Table 1. The Act enables the removal of any specified part from cadavers for therapeutic purposes, medical education or research under two conditions. First, it should be at the expressed request of the donor either in writing at any time or orally in the presence of two or more witnesses during the deceased's last illness. Second, it should be in the absence of an objection from the deceased and with the consent of the person lawfully in possession of the body (Kassim, 2005). The Act does not define the word "tissue" and also does not elaborate "the person lawfully in possession of the body". The Act does not provide a clear definition of death, and only requires two fully registered medical practitioners to confirm that life is extinct. Brain death is not mentioned in this Act although brain-dead donors are a source of organs in cadaveric organ transplantation. The Human Tissues Act 1974 only relates to cadaveric donors whereas live donors fall within the purview of the common law.

Since then, the Human Tissues Act has been amended to include brain death; it defines a method of determining death and it specifies that doctors who certify the brain death should not participate in the subsequent procedures for the removal and transplantation of the organs and tissues. The new Human Organ and Tissue Act will be tabled to the Parliament very soon. At present the MOH is drafting regulations to outline the details of the Act.

The Muzakarah (conference) of the Fatwa Committee of the National Council for Islamic Religious Affairs of Malaysia held on 21 June 1995 has permitted the use of tissue grafts namely amnion, bone and skin in medical practice, but tissues cannot be misused for trading or other purposes.

The National Organ, Tissue and Cell Transplantation Policy developed in 2009 is to guide practitioners in the field and all other stakeholders to further develop this therapeutic option to treat patients. The Policy is expected to promote further organ and tissue transplantation and those in need of such treatment will benefit from a comprehensive and efficient service. Again the Policy does not mention brain death and there are no details on tissues. The aims of the Policy are to ensure that organ, tissue and cell transplantation are carried out to the highest ethical and professional standards and to promote the highest quality of care including proper documentation and maintenance of registries. Confidentiality regarding the identity and personal details of donors and recipients must be ensured. The commercialisation of organ, tissue and cell transplantation and any acts that may indirectly promote or lead to commercial transactions are prohibited. Approved transplantations shall receive appropriate assistance from the government as there is a dedicated budget for the implementation of transplantation activities in the country. For efficient implementation, the National Transplantation Council (NTC) is appointed by the government as the governing body of the National Transplantation Programme to promote and monitor the progress of organ, tissue and cell transplantation in the country. A National Transplantation Technical Committee (NTTC) appointed by the Director General of Health is responsible for advising the NTC on matters related to policy. The NTTCC is also responsible for advising the National Transplantation Unit (NTU) on matters related to implementation. The NTU within the Medical Development Division of the Ministry of Health coordinates the implementation of organ, tissue and cell transplantation, and has responsibilities among others to

- Establish organ and tissue procurement units in hospitals;
- Develop and support organ, tissue and cell transplantation units in designated hospitals;
- Facilitate and support the development of national transplantation recipients, donors and donor pledgers registries;
- Implement training programmes for all personnel involved in organ, tissue and cell procurement and transplantation;
- Ensure the implementation of processes of the accreditation for organ, tissue and cell procurement and transplantation;
- Ensure that all practitioners in organ, tissue and cell transplantation are properly credentialed; and
- Promote public education activities.

The NTU is now introducing necessary or required amendments to existing legislation on transplantation. New regulations addressing the present shortcomings will soon be developed. For the past decade, there has been a steady increase in the rate of cadaveric organ donation (Muhammed Anis, 2012), with the highest number of cadaveric organ and tissue donors recorded in 2011. However, the cadaveric donation rate in Malaysia remains only 1.6 donors per million population (dpmp) which is far behind other countries such as Australia with 14.9 dpmp in the same year (Muhammed Anis, 2012).

In organising the cadaveric donation activity, a Transplantation Procurement Management Unit has been formed to coordinate organ and tissue procurement in all parts of the country with the local hospital unit managing the donor. Each identified hospital shall have a Tissue Organ

Procurement (TOP) Team consisting of trained personnel who shall be responsible for the identification and management of the potential donors including getting consent form the next-of-kin, evaluation for donation, and organising the procurement, storage and transport of the organs and tissues and speedy return of the donor's remains to the next-of-kin.

The policy clearly indicates that organs and/or tissues procured from donors in Malaysia shall not be allocated to recipients in another country unless there is no suitable recipient locally and there is a prior agreement on organ and/or tissue sharing between such a country and Malaysia. Importation of tissues from other countries shall be made through institutions recognised by the Ministry of Health and in accordance with the Guidelines on Importation and Exportation of Human Tissues and/or any Body Part.

National Transplantation Programme

All centres performing transplantation shall report to the National Transplant Registry (NTR). The report shall include details of the centre, the surgery and short- and long-term outcomes. All health professionals who are involved in transplantation should report to the NTR even though the data reporting is voluntary. Data should be collected systematically and in a standardised manner. The NTR website is made available for public reference (see Figure 7.2). The NTR shall report annually on all transplantation in the country. There shall be adequate provision for a comprehensive information and communication technology infrastructure and personnel to ensure efficient data management, record keeping, analysis, auditing, monitoring of outcome measures and research purposes. Research in all aspects of organ, tissue and cell transplantation is encouraged, including laboratory-based research, clinical trials, outcome and health economic studies. Research findings shall be made available at appropriate forums or publications in order to benefit the care of transplantation patients.

The NTR is a Ministry of Health (MOH)-supported registry whose aim is to collect information about organ and tissue transplantations in Malaysia. The NTR collects information about patients who have undergone

Figure 7.2. The National Transplant Registry website.

organ or tissue transplantation. The information allows an estimation of the magnitude of transplant activity in the country. Such information, besides being useful to transplantation practitioners, is also useful in assisting the MOH, non-governmental organisations, private providers and the industry in programme planning and evaluation of transplantation services.

The objectives of the NTR are to

- Determine the frequency and distribution of all types of transplantation activity in Malaysia;
- Determine the outcomes of transplantation;
- Determine the factors influencing outcomes of transplantation;
- Evaluate transplantation services in the country;
- Stimulate and facilitate research on transplantation and its management.

The NTR is co-sponsored by the following organisations of the MOH:

- National Transplant Coordinating Committee, MOH
- Medical Development Division, MOH
- Malaysian Society of Transplantation
- Clinical Research Centre (CRC), Kuala Lumpur Hospital, MOH

Current legislation allows doctors to release their patients' data to NTR as the data is essential to public health and safety. However, the NTR has also developed strict information security policies and procedures to protect data confidentiality in accordance with standard disease registration practice and in compliance with professional standard and applicable regulatory requirements.

The NTR receives data on organ/tissue transplantation from the following three main sources:

1. The individual doctors who provide transplantation services, who voluntarily report data to the NTR. Data collection will be from seven main types of transplantation services:

    - Blood and marrow transplant;
    - Cornea transplant;
    - Heart and lung transplant;
    - Liver transplant;
    - Renal transplant;
    - Heart valve transplant;
    - Bone and tissue transplant.

2. The National Vital Registration System (National Registration Department of Malaysia). Their data is useful for determining or verifying mortality outcomes of transplant patients.
3. Information Documentation Unit of the MOH, which operates the Health Management Information System (HMIS).

The NTR has published annual reports which are made available from the NTR website (see Figure 7.3).

The National Transplant Resource Centre (NTRC) was established by the MOH in the early 2000s with the support and cooperation of the Medical Development Division, MOH. The main task of the NTRC is to spread correct information and increase awareness about organ and tissue donation and transplantation in Malaysia.

The NTRC is responsible for getting more individuals to pledge donations, and it acts as the registrar for organ, tissue and cell donation in Malaysia. The NTRC collaborates with various non-governmental organisations, religious bodies, students' associations, corporate bodies and the media to disseminate information on donation and transplantation to the public. The NTRC personnel are known as "transplant coordinators" who

Figure 7.3. Seventh National Transplant Registry Annual Report 2010.

have roles in identifying potential donors, approaching the family to advise on organ and tissue donation, alerting all the recipient teams, getting written consent from the family, managing all the necessary setting such as police and magistrate requirements, coordinating all various teams and individuals for procurement, ensuring the availability of operating theatres, monitoring the whole process of procurement and finally, returning the body back to the family for the funeral.

Certification and Compliance

At present only the USM Tissue Bank has been certified. The USM Tissue Bank with good management practice (GMP) set up has been certified to be compliant with the following:

- Quality Management System ISO 9001:2008 (not compulsory)
- Conformance to GMP for the bank facility
- MS ISO 13485:2006 (Medical Device)
- MS ISO 1900:2005 (Shariah Perspective)

Quality Management System ISO 9001:2008 (not compulsory)

Quality Management System (QMS) is a management system to direct and control an organisation with regard to quality. The certification is on a voluntarily basis but is highly encouraged. The eight Quality Management Principles are: Customer Focus, Leadership, Involvement of People, Process Approach, System Approach to Management, Continual Improvement, Factual Approach to Decision Making and Mutual Beneficial Supplier Relationship. To develop and implement an effective QMS, a tissue bank must

- Determine needs and expectations of customers;
- Establish quality policy and quality objectives;
- Determine needed processes and responsibilities;
- Provide resources to attain quality objectives;
- Establish methods for process measurement;
- Determine process efficiency and effectiveness;
- Determine means to prevent non-conformities;
- Apply process for continual improvement.

The following are documents to be established:

- Quality manual — quality policy, organisation structure, responsibilities and authorities, training, audit, management review;
- Standard operating procedures (SOPs) — manufacturing processes, equipment management, environment, quality control;
- Work instruction;
- Records and reports.

The bank should have a comprehensive Quality Assurance (QA) system incorporating GMP. The QA system should be fully documented and its effectiveness should be monitored. All parts of the QA system should be adequately resourced with

- competent personnel; and
- suitable and sufficient premises, equipment and facilities.

The QA system

- ensures that tissue allografts are released after review by an authorised person;
- provides satisfactory arrangement to ensure tissue allografts are stored, dispatched and handled appropriately;
- places a mechanism for regular self-inspection/internal quality audit.

Conformance to GMP for the bank facility

Good Manufacturing Practice (GMP) requires that manufacturers of regulated healthcare products follow this Practice to ensure the safety, identity, strength, purity and quality of the tissues. GMP is a part of QA which ensures that tissue allografts are consistently processed and controlled to the quality standards appropriate to their intended use. GMP can

- Minimise risks of cross contamination and mix-up;
- Ensure tissues/materials are traceable to the original source; and
- Ensure that quality is built into the product.

In Malaysia, inspection of manufacturing premises and licensing and issuance of certificates is under the Centre for Compliance and Licensing, National Pharmaceutical Control Bureau (NPCB) of MOH. At present, cell and tissue products are still not considered as registered products. However, it is within the programme of the future NPCB regulatory framework. Currently the GMP inspection conducted at cell and tissue manufacturing/processing facilities are voluntary. Nevertheless, the tissue banks which comply with GMP requirements will be issued a manufacturer's licence. In Malaysia, this is referred to the Pharmaceutical Inspection Cooperation Scheme (PIC/S) GMP Guidelines. Malaysia has been one of the 35 countries participating in the PIC/S since 2009. Quality Risk Management (QRM), a guideline developed by the International Conference on Harmonisation of Technical Requirement (ICH) has recently been included as a new annex to PIC/S. QRM is a systematic process for the assessment, control, communication and review of risks to the quality of the product. The QRM system should ensure that

- The evaluation of the risk to quality is based on scientific knowledge, experience with the process and ultimately links to the protection of the patients; and
- The level of effort, formality and documentation of the QRM process is commensurate with the level of risk.

In line with the awareness and importance of biovigilance in tissue banking, QRM is considered the best approach. Records must be accurate, complete, legible and indelible. Accurate information is necessary to maintain traceability throughout the chain of processing and dispatch/distribution. Traceability is the ability to locate and identify the tissue/cell during any step — from procurement, throughout processing, testing and storage, to distribution to the recipient or disposal. This implies the ability to identify the donor and the tissue establishment or tissue bank facility receiving, processing or storing the tissues/cells and the ability to identify the recipient(s) at the medical facility. Traceability also covers the ability to locate and identify all relevant data relating to products and materials coming into contact with those tissues/cells.

Quality Control (QC) is part of GMP. QC is concerned with sampling, specification and testing. The QC unit should be independent from the production unit. The scope of QC in tissue banking concerns the following:

- Procured tissues — medical screening and laboratory tests;
- Starting materials — compliance with specifications;
- Intermediate products — bioburden prior to sterilisation and sterilisation records;
- Tissue allografts as finished products — packaging and labelling; and
- Environmental conditions — temperature, humidity.

MS ISO 13485:2006 (Medical device)

The status of tissue allografts as a product in Malaysia is still not clear. However xenografts obtained from cows and subjected to aggressive processing by the USM Tissue Bank are considered to be a medical device. Tissue banks in Malaysia have to submit product information and specifications to the Medical Device Authority (MDA) for evaluation. The MDA will decide if any of the tissue products need to be registered under the MDA. One of the functions of the MDA is to implement, enforce, consider and recommend reform to the medical device laws.

A medical device regulatory system has been implemented in Malaysia since 2012 with the objectives being to

1. Address public health and safety issues by
    - Conducting pre-market control to assess safety, effectiveness and quality of medical devices;
    - Providing adequate information for the public and health professionals to make informed choices on medical devices;
    - Controlling the usage of certain medical devices;
    - Providing a post-market reporting system to identify and monitor medical devices with problems in the market.
2. Facilitate the medical device trade and industry by
    - Encouraging local manufacturers to market their products globally;
    - Providing a favourable environment for the growth of the medical device industry.

There are two Acts with regard to the medical device regulatory system (Laws of Malaysia, 2012):

- Medical Device Authority Act 2012 (Act 738); and
- Medical Device Act 2012 (Act 737) and subsidiary regulations.

The Medical Device Act (MDA) 2012 was gazetted in February 2012 and enforced on 30 June 2013. The Medical Device Regulation (MDR) 2012 was gazetted in December 2012 and was effective since 1 July 2012.

Under the MDA 2012 (Act 737) all medical products shall be registered prior to placement into the market. All establishments shall be licensed to conduct activities relating to manufacturing, importing, representing foreign manufacturer and distributing medical devices in Malaysia. A medical device is defined as a product which is used for human medical purposes. If applied to the body, the effect of the medical device is primarily physical. The product includes any instrument, apparatus, implement, machine, appliance implant, *in vitro* reagent or calibrator, software, material or other similar or related article. The definition seems not to cover tissues and tissue products; however, Section 3(1) of Act 737 states that the establishment shall classify its medical device based on the level of risk it poses, its intended use and vulnerability of the human body in accordance with the prescribed manner. MDR 2012 indicates that the level of regulatory control is proportional to the risk associated with the medical device (risk-based classification). Products are classified based on rules in the First Schedule (rules of classification of medical device) and grouped using the rules in Second Schedule of the Act.

Conformity Assessment (CA) shall be conducted to provide objective evidence of safety and performance of a medical device and compliance to regulations. Four elements of CA are Quality Management System (QMS), Post-market Surveillance System (PMS), Summary Technical Documentation and Declaration of Conformity (DoC). Post-market surveillance and vigilance are important to deal with adverse reactions towards any products. If there are such complaints, tissue banks must know what action to take, how to handle the problem and how to report it.

MS ISO 1900:2005 (Shariah perspective)

This is a QMS which includes requirements from the Islamic perspective. USM Tissue Bank has recently been certified to this QMS, MS 1900: 2005 (Firdaus and Suzina, 2014). It is a structured system to assist an organisation to deliver products and services meeting customers' requirements, and be compliant with regulatory requirements as well as the requirements of Shariah. It is based on ISO 9001:2008 but includes the requirements of Shariah, i.e., it practises the Islamic values system in daily work and inculcates Islamic values system as the foundation for customer satisfaction and organisational performance. According to Islamic values, the bank should deliver the best available products/services and on time to the customers. The main aim is to enhance Muslim customers' satisfaction, and to give assurance of conformity to applicable regulatory requirements and assurance of compliance to Shariah requirements. Shariah is the total of Allah's commands and Sunnah is that which regulates all aspects of the life of every Muslim. Therefore the certification covers all core activities, namely, tissue procurement, processing and distribution as well as the supporting activities, i.e., financial, training and infrastructure. Examples of Shariah requirements in each of these activities include the following:

- Source of tissue: for bovine tissue, the tissue must be halal-certified;
- Processing: all chemicals used must be halal and clean;
- Distribution: source of finance must be halal, free from any *riba* (usury), and any agreement must be fulfilled.

USM Tissue Bank is not only the first bank in Malaysia; in fact, it is the first tissue bank in the world that complies to this certification whereby the bank is managed in accordance with universally accepted values such as justice, honesty, truthfulness, sincerity, timeliness and discipline and that the halal and non-halal aspects of all processes necessary for the delivery of services and products are identified and communicated within the organization and effectively implemented. Under Shariah Critical Control Points (SCCP), every activity must be assessed and confirmed in its conformance to Shariah. The tissue bank has a competitive edge in convincing the Muslim customers, especially in accepting bovine-based grafts.

Routine operations

For routine operations, the tissue banks in Malaysia refer to tissue bank standards established by the American Association of Tissue Banks (AATB), Asia Pacific Association of Surgical Tissue Banks (APASTB) and IAEA. The following documents are also referred to for validation work and in establishing standard operating procedures for tissue banks:

- ISO 11137:2006 Sterilization of healthcare products
- ISO 11737:1995 Sterilization of medical devices — Microbiological methods
- IAEA Code of Practice 2007: Radiation sterilization of tissue allografts: Requirements for validation and routine control.

CONCLUSION

Although tissue banking has been taking place in the Asia-Pacific region for more than 30 years, the introduction of legislations is generally rather slow. To ensure proper management and functioning of a tissue bank, the legal status on procurement, banking and transplantation is urgently needed and must be clarified. Undoubtedly, religion and cultural practices are important factors in establishing the laws as these aspects are very important to achieve effective tissue donation which is the essential factor for any tissue bank to sustain its function to serve the country. The Acts and Regulations described here can ensure that the supply of safe tissue allografts will steadily increase with the demands to fulfil the needs of patients. Experience and challenges faced by the Malaysian tissue banks in meeting national legal requirements and needs for certification are shared here.

ACKNOWLEDGEMENTS

The author would like to thank Professor Thin Lwin of Tissue Bank, Yangon Orthopaedic Hospital, Myanmar; Associate Professor Suzina Sheikh Abdul Hamid of USM Tissue Bank, Malaysia; and Mr Hemaka De Mel of Human Tissue Bank, Sri Lanka; for providing the current legal status of tissue banking in the respective countries.

REFERENCES

Bagheri A (2005) Organ transplantation laws in Asian countries: A comparative study, *Transplant Proc* **37**, 4159–4162.

Bile KM, Qureshi JA, Rizvi SA, Naqvi SA, Usmani AQ, Lashari KA (2010) Human organ and tissue transplantation in Pakistan: When a regulation makes a difference, *East Mediterr Health J* **16**(Suppl.), S159–S166.

Ding C (2008) Latest development of legal regulations of organ transplant in China, *J Int Bioethique* **19**(4), 61–81.

Firdaus AR, Suzina SAH (2014) Quality Management System from Islamic Perspectives: USM Tissue Bank Experience, Melbourne, Australia, 12–14 May 2014.

Gajiwala AL (2003) Setting up a tissue bank in India: The Tata Memorial Hospital experience, *Cell Tissue Bank* **4**, 193–201.

Hilmy N, Manjas M, Abbas FB, Morales Pedraza, J (2009) The impact of the International Atomic Energy Agency (IAEA) Program on Radiation and Tissue Banking in Indonesia, *Cell Tissue Bank* **10**, 103–107.

Ireland L, McKelvie H (2003) Tissue Banking in Australia, *Cell Tissue Bank* **4**, 151–156.

Joo HN (2013) The Organ Transplantation Act and recent trends in Korea, *Asia Pac J Public Health* **25**(2), 209–213.

Kassim PN (2005) Organ transplantation in Malaysia: A need for a comprehensive legal regime, *Med Law* **24**(1), 173–189.

Laws of Malaysia (1974) Human Tissues Act 1974: Act 130.

Laws of Malaysia (2012) Medical Device Authority Act 2012: Act 738.

Laws of Malaysia (2012) Medical Device Act 2012: Act 737.

Lee SD, Kim, JH (2009) Changes in the organ procurement system in South Korea: Effects on brain-dead donor numbers, *Transplant Proc* **41**(9), 3551–3555.

Ministry of Health Malaysia (2009) National Organ, Tissue and Cell Transplantation Policy.

Morales J, Phllips GO (2004) IAEA: Past and present, in Phillips GO, Strong DM, von Versen R, Nather A (eds.) *Advances in Tissue Banking*, Vol. 7, World Scientific, Singapore, pp. 3–12.

Muhammed Anis AW (2012) Brain Death Audits, 15th Scientific Meeting of the Malaysian Society of Transplantation, Kuching, Malaysia, 17–19 May 2012.

Nather A, Low JM, Sim, YL (2010) Ethical, religious, legal and cultural issues for tissue banking, in Nather A, Yusof N, Hilmy N (eds.) *Allograft Procurement, Processing and Transplantation: A Comprehensive Guide for Tissue Banks*, World Scientific, Singapore, pp. 107–120.

Nather A, Zheng S (2010) Evolution of allograft transplantation, in Nather A, Yusof N, Hilmy N (eds.) *Allograft Procurement, Processing and Transplantation: A Comprehensive Guide for Tissue Banks*, World Scientific, Singapore, pp. 3–42.

Ohkawara H, Fukushima N, Kitagawa T, Ito T, Masutani Y, Sawa Y (2010) Tissue procurement system in Japan: The role of a tissue bank in medical center for translational research, Osaka University Hospital, *Transplant Proc* **42**(1), 190–192.

Phillips GO, Jorge Morales Pedraza (2013) The use of the ionization technique for tissue sterilisation: The International Atomic Energy Agency (IAEA) experience, *Cell Tissue Bank* **14**, 341–348.

Republik Indonesia (2009) Undang-undang Nomor 36 Tahun 2009: Tentang Kesehatan.

Shroff S (2005) Cadaver Organ Donation and Transplantation in Asia: The Way Ahead, Fourth International Congress of Kidney Diseases, Ahmedabad, 11–13 November 2005.

The Gazette of India (2014) The Transplantation of Human Organs and Tissue Rules, Ministry of Health and Family Welfare Notification, New Delhi, 27 March 2014.

The Gazette of Pakistan (2010) The Transplantation of Human Organs and Tissue Act: Act No. VI of 2010, 18 March 2010.

The Union of Myanmar (2004) The Body Organ Donation Law (The State Peace and Development Council Law No.1/2004), 19 February 2004.

World Health Organisation (2010) Guiding Principles on Human Cell, Tissue and Organ Transplantation, EB123/5.

Zahid HM, Chakma K, Miah, M, Nessa, A (2010) Ethical perception of tissue banking in Bangladesh, *Bangladesh J of Bioethics* **1**(2), 11–19.

Zheng MH, Pembrey R, Niutta S, Stewart-Richardson P, Farrugia A (2006) Challenges in the evaluation of safety and efficacy of human tissue and cell-based products, *ANZ J Surg* **76**, 843–849.

8

LEGAL SYSTEM FOR TISSUE BANKING IN LATIN AMERICA

Eulogia Kairiyama

INTRODUCTION

Organ transplantation is an established effective therapy that saves lives and significantly improves the quality of life. Over the last couple of decades, the donation and transplant of human organs, tissues, and cells has progressed rapidly but irregularly among the Latin America countries. These countries' health ministries, together with regional, interregional, and international organizations have contributed to diminish these differences. These contributions were mainly due to initiatives from the International Atomic Energy Agency (IAEA) and the Organización Nacional de Trasplantes (ONT) (Spanish National Transplant Organization), which have major influence in Latin America on the establishment, development, training of human resources, and consolidation in several countries with regard to the donation, transplantation, and banking of organs, tissues, and cells.

Through the powerful infrastructure supported by the IAEA Technical Cooperation Program, several national, regional, and interregional projects were implemented, all aimed to ensure a great scientific and technological development in Latin America and other international regions (Morales and Phillips, 2008; Morales and Phillips, 2009; Nather and Morales, 2009; Phillips and Pedraza, 2003).

In Latin America 12 countries have benefited through national, regional, and interregional projects from 1993 to 2012 (Morales and Phillips, 2008; Kairiyama and Morales, 2009; Herson *et al.*, 2009; Aguirre and Morales, 2009; Morales *et al.*, 2009; Martínez *et al.*, 2009; Castro Gamero and Morales, 2009; Alvarez *et al.*, 2009). The impact of the IAEA Program on Tissue Banking and Radiation Sterilization in Latin America can be summarized as follows:

- Establishment of tissue banks using ionising radiation for sterilization;
- Increase of the quality of the production of tissue allografts;
- Increase in the production of sterilized tissues using ionising radiation for medical treatment;
- Reduction in the import of tissues by increasing the local production of tissues;
- Sustainability in the number of donors through the implementation of a public and professional awareness campaign;
- Training of tissue bank operators and medical personnel from Latin American countries through regional and interregional training courses as well as through individual training in Europe, Asia-Pacific countries, and the United States. Argentina has been selected by the IAEA to host the Regional Training Center for the Latin American region;
- Implementation of regional training courses in Buenos Aires, through both face-to-face and virtual modalities;
- Training course curriculum preparation in English by Asia-Pacific countries, and its update and translation to Spanish for the Latin America region, to train tissue banks operators;
- Publication of several materials related to tissue banking and radiation sterilization of tissue allografts, codes of practice for radiation sterilization of human tissues, and recommendations for the operation of tissue banks.

To ensure an appropriate and safe use of the radiation sterilization technique for the sterilisation of tissues, the IAEA, within the framework of its Program on Radiation and Tissue Banking, elaborated the document entitled *Radiation Sterilization of Tissue Allografts: Requirements for Validation and Routine Control: A Code of Practice* (IAEA, 2007). This Code of Practice adopted the principles of the International Organization for Standardization, the ISO 11137:1996 for Radiation Sterilization of Health Care Products (ISO 11137:1995). The approach has been adapted to take into account the special features associated with human tissues and the features that distinguish them from industrially produced sterile healthcare products. The approach, as described here, is not applicable if viral contamination is identified. Thus it is emphasized that the human donors of the tissues must be clinically and serologically screened.

Through the Latin America Regional IAEA Project (ARCAL CVIII, RLA 6062), the abovementioned Code of Practice published in 2007 was reviewed, translated, updated (ISO 11137:2006) and published as *Código de Prácticas para la Esterilización de Tejidos para Injerto con Radiación Ionizante* (Kairiyama, 2014) as well as the document *Guía para la Operación de Bancos de Tejidos* (Alvarez, 2014), a general recommendation for tissue banks. Both these documents were published in March 2014 by the Tecnológica Costa Rica Editorial (TCR).

HISTORY OF TISSUE BANKING AND TRANSPLANTATION IN LATIN AMERICA

In Latin America, as in all over the world, there is a great interest in improving the health and quality of life of patients by organ and tissue transplantation. From ancient times Man has always been interested in replacing damaged parts of the body, deformed or sick. The knowledge of transplants goes back to the Aztec empire, when bone grafts were carried out on bone fractures (Sahagún, 1955).

In Mexico, in the modern era, the history of transplants in the country started in 1870 when Dr Luis Muñoz published the first data about a successful skin graft performed at San Andrés Hospital. The first bone and corneal transplants were performed in 1919 and 1940, respectively. Tissue banking as we know it today was started in Mexico when a bone bank was established at the Children´s Hospital of Mexico in 1948. Another bone bank was set up at the Central Military Hospital in 1949. The third bone bank started its activities in 1950 at Juárez Hospital. Other bone banks were established at Primavera Clinic, Rubén Leñero Hospital, and Xoco Hospital, all in Mexico City. However, these banks were canceled or stopped for several reasons (Castellanos, 1985).

It is noteworthy that the first attempts to transplant corneal tissue were performed in 1890, in Sucre city, Bolivia (Instituto Nacional de Oftalmología). The protagonists were Dr Gerardo Vaca Guzman Moyano, considered the creator of specialty ophthalmology in Bolivia, and José Manuel Camó and Montovio, Spanish ophthalmologists who pioneered this specialty.

In Argentina, Dr Antonio Manes and his team performed the first transplant of corneas at Rawson Hospital, Buenos Aires, in 1928, and Dr Carlos E. Ottolenghi performed the first bone transplant at the Italian Hospital, Buenos Aires, from a cadaveric bone piece banked in a frozen state in 1948 (INCUCAI-Historia; Ottolenghi, 1966).

In Uruguay, the healthcare of patients with a clinical therapeutic necessity for tissue and organ transplantation was attended to at the Clínicas Hospital of the Faculty of Medicine of the University of the Republic, with the first corneal transplantation in 1946 and the development of clinical application of vascular allografts in the 1950s. While clinic transplantation activities were developed, the legal framework in donation and transplantation in this country was implemented with the support of the lawyers Dr Adela Reta and Dr Bayardo Bengoa (Álvarez and Saldías, 2013).

The early history of conservation of human skin for transplantation in Brazil dates back to 1956, which was performed at the Plastic Surgery Department of the Hospital das Clinicas, Sao Paulo (Herson *et al.*, 2009), where it was stored refrigerated for a short time (nowadays skin is preserved in more than 75% glycerol).

The first multi-tissue bank of Cuba was founded in Havana in 1958. At that time, freeze-drying was used at the bank as a method of preserving bone, as well as irradiation with Cobalt -60 (^{60}Co), the radiation source to sterilize bone, heart valves, and other tissues (Morales *et al.*, 2009).

SCOPE OF REGIONAL INTEGRATION

In order to progress the development of transplants in Latin America and the Caribbean, health organizations have signed several agreements among Latin American countries and also with the Iberian Peninsula (Spain and Portugal). They have created a regional strategy to promote the donation and transplantation of organs, tissues, and cells (INCUCAI-Integración regional).

The following groups and networks have been set up:

- Grupo Punta Cana. This group emerged in 2001. The Latin American countries are linked to this group and adopted the Spanish Model as a guide for national procurement programs.
- Red Consejo Iberoamericano de Donación y Trasplantes (RCIDT) (Network Council of Donation and Transplant). In 2005, the Latin America Ministries of Health together with Spain and Portugal created an institution, the RCIDT, in order to develop and improve cooperation among Latin American countries in the organizational and legislative aspects, training professionals, and ethical and social issues related to donation and transplantation of organs, tissues, and cells (Transplant Observatory on Donation and Transplantation, RCIDT). Each Latin American country is represented by a professional in the system for transplants and/or organization, nominated by the corresponding Minister of Health. Moreover, the Organización Nacional de Trasplantes (ONT) (Spanish National Transplant Organization) was chosen to be the Permanent Secretariat of the RCIDT. The first meeting of the Council of Donation and Transplant was carried out in Mar del Plata, Argentina, in November 2005, where its members signed the Mar del Plata Statement. This Statement makes recommendations on ethical, legal, technical, and organizational aspects that member states should reach in their countries regarding the donation and transplantation of organs, tissues, and cells of human origin. It considers as prerequisites the safety, quality, efficacy, and ethics in the practice of transplantation and the existence of national legal norms to regulate the activity of donation and transplantation of organs, tissues, and cells; subordinate to a national official organization as regulator and coordinator of the activity under the National Health Authority. Transparency in transplantation activities at the national level is essential for responsibility and traceability, and to prevent commercialization of organs, tissues, and cells. It also recommends to member states to introduce or revise regulatory standards, and use or adapt existing laws, regulations, documents, and definitions in the field of transplantation of organs, tissues,

and cells, which are accepted and agreed by international bodies. National Regulatory Standards should ensure at least the following principles: altruism in the donation; national organ and tissue allocation in living donations limited to related family and avoiding unrelated living donations; waiting lists access and adherence to minimum standards of quality and safety of the biological implant material (Declaración de Mar del Plata, 2005). In addition, note should be taken of the Spanish ONT program called the Master Alliance in Donation and Transplantation of Organs, Tissues and Cells, whose main objective is to move to the Latin American countries the successful Spanish Model of coordination through training of professionals in Spanish centers and specialized courses (ONT-EL Modelo Español, 2008). At the Mar del Plata meeting the authorities of the Council of Donation and Transplants were chosen. Rafael Matesanz, Head of the ONT, and Carlos Soratti, President of INCUCAI, were unanimously elected President and Vice-President, respectively.

Since the creation of the RCIDT, over ten meetings have been held in most countries of the region, in which recommendations were made and various documents approved, which are published in the newsletter *Transplantation Latin America*.

- The Council on Donation and Transplantation of MERCOSUR. Argentina, Brazil, Paraguay, and Uruguay, countries that conform with MERCOSUR (South Common Market), developed increased levels of integration to ensure equitable access of the population to transplantation of organs, tissues, and cells.
- The Comité Intergubernamental de Donación y Trasplante (CIDT) of MERCOSUR (Intergovernmental Committee on Donation and Transplantation). This was set up by those responsible for the areas of donation and transplantation in Brazil, Venezuela, Uruguay, and Argentina, as permanent members, and other Latin American countries as Associated States. It was created in 2006 in order to promote cooperation in the training of human resources, development of analysis of the economic aspects of transplantation, and the establishment of common standards of quality and safety in procurement, processing, preservation, distribution transportation, and implantation of organs, tissues, and cells.

- In 2011, the Intergovernmental Committee on Donation and Transplantation (CIDT) set up the Regional Registration Donation and Organ Transplant (DONASUR), which was approved by the National Health representatives of the region. The MERCOSUR Donation and Transplantation Registry was proposed by Argentina; in addition, countries that are currently involved are Bolivia, Brazil, Chile, Colombia, Ecuador, Paraguay, Peru, Uruguay, and Venezuela. It enables comparison between countries, unifies languages and criteria, and ensures transparency and process safety. DONASUR also includes regional lists of emergency.
- The Association of Universities Grupo Montevideo brings together representatives from universities in Argentina, Bolivia, Brazil, Chile, Paraguay, and Uruguay, and includes a Core Discipline Donation and Transplantation Group. The Association aims to contribute to the strengthening and consolidation of high-level human resources, building on the comparative advantages of installed capacity in the region in scientific and technological research, continuing education, and interaction with members of their society as overall, disseminating advances in knowledge which fosters modernization.

CURRENT TISSUE BANKING ACTIVITY

The legal framework allows implementing an effective national oversight on all the activities of donation and transplantation of human organs, tissues, and cells; from the donation of the human material for transplantation up to the recipient. This also allows the traceability, safety, and effectiveness of the material applied, equitably meeting the transplantation needs of the patient, transparency of the organization, and execution of the donation and transplantation activities, and prevents the sale of human material (see Figure 8.1).

In order to produce and distribute safe and clinically efficient tissues for transplantation to a patient, all the steps of tissue banking activity regarding tissues and cells, from the selection of the donor up to transplantation in a patient, are regulated and implemented by the National Health Authority. It includes ensuring accountability of the human material for

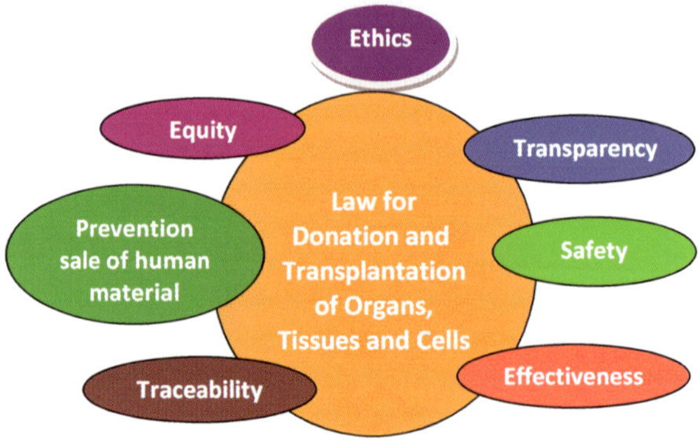

Figure 8.1. Legal framework of organ, tissue, and cells banking activity scheme.

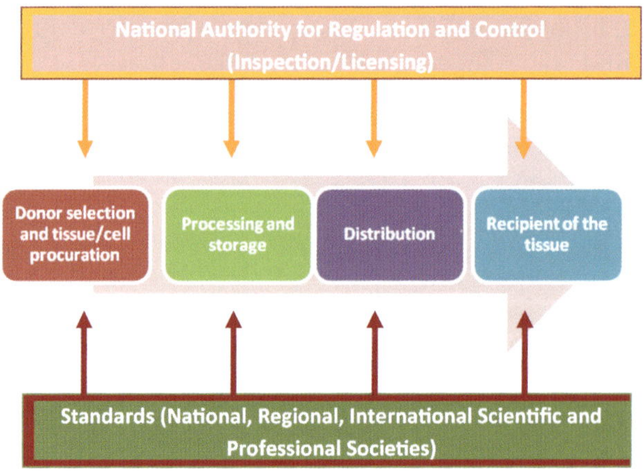

Figure 8.2. Process of production of tissues and cells for implantation scheme.

transplantation and its traceability in the frame of the country's laws and legislations.

This activity is supported by standards provided by national, regional, or international scientific and professional societies in order to meet the requirements of each regulation under a quality system (see Figure 8.2).

The Pan America Health Organization (PAHO) (OPS, 2013) examined the legislations in the Latin American countries and found that the regulatory development of the Latin American region related with donations and transplantation of organ, tissues, and cells varies significantly from country to country. So, on the one hand the legislation is complete, updated, and recent, and on the other it lacks specific standardization, and is different from that enunciated in the general laws of health policy. Many current regulatory frameworks date back several years ago, which means that in some cases technological advances are not included therein, or need to be regulated by further decrees or resolutions. These uncorrelated situations create difficulties when establishing which are the current regulations for a particular aspect. The general contents of the law are similar across the countries. However, there are differences in the detail and precision of how these are regulated.

OVERVIEW OF LEGISLATION AND THE CURRENT SITUATION REGARDING TISSUE BANKING IN LATIN AMERICA

The following countries conform to varying degrees: Argentina, Bolivia, Brazil, Chile, Colombia, Costa Rica, Cuba, Dominican Republic, Ecuador, El Salvador, Guatemala, Honduras, Mexico, Nicaragua, Panama, Paraguay, Peru, Uruguay, and Venezuela.

Argentina

Due to the growing demand for transplants, the Argentine government was faced with the need to regulate the practice. Therefore, in 1977, it enacted Law 21541 that gave birth to the Centro Único Coordinador de Ablación e Implante (CUCAI) (Unique Coordinator Center of Ablation and Implant). The CUCAI, which began operating in 1978, emerged as the national procurement agency in charge of standardization of the activity (INCUCAI).

In Argentina the donation of organs and tissues is based on solidarity by a voluntary and altruistic act. It is a confidential action that depends on the individual will, whose objective is the common well-being. The Transplant of Organ and Tissue Law No. 24193/1993 was later amended and replaced by Law No. 26066 in 2005. The removal of organs and tissues for implantation is regulated by the requirements of this law throughout all the Argentina Republic. Articles 9 and 10 states that the medical procedures can only be carried out in medical establishments registered with the respective jurisdictional authority controller. The inscription shall be valid for a period not exceeding two years. Renewal can only be made after (1) inspection of the facility by the jurisdictional authority controller in all cases as a requirement for such registration; (2) sufficient evidence that the establishment has adequate physical infrastructure and instrumentation, and (3) the facility has the necessary qualified personnel in the specialty. The naturally renewable and separable body tissues are excluded, with the exception of procuring and preserving of hematopoietic progenitor cells (CPH) and subsequent implantation in humans, which shall be regulated by this law. This law incorporated the alleged donor for organ donation, and states that everyone over 18 years old is considered a donor, except those who state the contrary. It came into effect in 2006 and allowed procurement of organs and tissues to continue increasing. In 2012, Argentina reached a rate of 15.7 donors per million inhabitants, positioning it at the forefront of the region's donation (INCUCAI).

In 1990, Law 23885, a modification of Law 21541, established the National Institute Unique Central Coordinator of Ablation and Implant (INCUCAI). The INCUCAI is a decentralized entity under the Minister of the Ministry of Health. Its actions are oriented to the application of the Law on Organ and Tissue Transplantation. It is the body that promotes, standardizes, coordinates, and supervises the activities of donation and transplantation of organs, tissues, and cells. It works in 24 Argentine Provinces with jurisdictional agencies of ablation and implantation in order to give people a transparent and equitable access to transplantation. INCUCAI assumes specific responsibilities assigned to it by the legislation, among which the following are included: it establishes and continuously updates the waiting lists of potential recipients; it coordinates the distribution of organs and tissues for transplantation; it determines procedures for maintenance of potential donors and transport organs; it makes rules for the authorization of establishments that carry out medical acts and the authorization of the professionals who practice it.

In 1993, the INCUCAI launched the Federal Program of Organ and Tissue Procurement, with the basic premise of increasing organ procurement. In that year the Registro Nacional de Células Progenitoras

Hematopoyéticas (CPH) (National Donor Registry of Hematopoietic Progenitor Cells), created by Law 25392, began to work in the field of INCUCAI to respond to people requiring CPH transplants, who did not have a matched donor within their families.

In 2010, the Ministry of Health through the INCUCAI and the procurement system implemented the Hospital Donor Strategy, which aims to incorporate the procurement as an activity of hospitals to increase the availability of organs and tissues for transplantation.

Currently, the Argentine health system has trained thousands of professionals in this field and the INCUCAI coordinates and supervises enabled centers for organ transplant (149 centers) and for tissue and cells (more than 600 centers), throughout the country (INCUCAI).

INCUCAI resolutions related to the activities of tissue banking and tissue transplantation are the following (INCUCAI):

- Resolution 069/2013: Norms and procedures for the registration of procurement activity, processing, storage, distribution, and implant tissues from living donors and cadaveric donors except ocular tissues.
- Resolution 118/2009: Standards and procedures pertaining to the activities of tissue banks. Scope: requirements for bank facility, human resources, standard operating procedure manual, additional services, sterility certification, technical and administrative procedures, and distribution in the process of donation and ethical rules of tissue banks. The resolution includes Annex VI — Good Tissue Practices for Transplantation. This standard is aimed at ensuring that human tissues are safe and effective, and establishes requirements for the procurement, processing, storage, distribution, and transportation of human tissues for transplantation. Each tissue bank shall establish and maintain a quality system to ensure that the requirements of this standard are met and that the produced tissues are safe, efficient, and effective, and appropriate to the intended use. The requirements of this standard shall apply to any human tissue to be procured or imported for implantation in the whole of the country.
- Resolution 001/2011: Norms and registration of authorized establishments: musculoskeletal and osteoarticular system, eye banks, valvular and vascular banks, and skin banks. Norms and registration of professionals accredited to implant tissues.

- Resolution 119/2012: Standards for good manufacturing practices and laboratory cell preparations.
- Other related resolutions numbers: 271/2009; 187/2001; 148/2001; 319/2004.

The elaboration of the technical principles of rules or updates is carried out by the participation of Advisory Committees. There are eight active commissions coordinated by the INCUCAI Scientific Technical Management regarding organ transplantation (seven committees) and tissue banking (one committee).

At the end of 2005 the INCUCAI implemented the National System Information of Procurement and Transplantation in Argentina (SINTRA). It is a computer system for the administration, management, supervision, and consultation activity procurement and transplantation of organs, tissues, and cells at the national level. An integrated information system with jurisdictional, regional, and national scope, allows real-time recording of the activity, the management of patients on dialysis, waiting lists, and allocation of organs and tissue for implantation in the country, with steps that enable ongoing monitoring and evaluation, as well as offering the society ensured transparency of the activity (INCUCAI-SINTRA).

The concept of Quality Assurance in the donation–transplantation process ensures the implementation of uniform records throughout the country, allowing the monitoring of operational steps for the detection of potential donors up to transplantation to the recipient. These records are managed in real time by the INCUCAI computer system, the SINTRA, which allows access to information with different levels of complexity. Using the SINTRA, it is possible to see the continuous and coordinated work of INCUCAI, jurisdictional agencies, and transplant centers.

The following paragraphs outline some milestones in the development of the tissue banking activity in Argentina.

The first Skin Bank was established in 1989 at the Fortunato Benaim Burn Foundation (Bolgiani, 1991). The skin was stored in frozen condition and gamma radiation was used from 1993 to sterilize the tissue. The first experimental research on radiation sterilization of human tissues and the development of Good Tissue Irradiation Practice was carried out in the National Commission of Atomic Energy (CNEA), at the Ezeiza Atomic Center (Kairiyama *et al.*, 1994).

Also at the Burn Foundation, the first laboratory of epidermis culture was installed in 1992. It allowed the treatment of many severely burned patients by the team of Dr Fortunato Benaim, as well as in many burn centers in Buenos Aires (Lorenti, 1991). Also, cultured skin was applied to patients after resection of the giant nevus, or to treat scars from previous burns. It was even used in patients with irradiated skin from a tissue bank; the homologous dermis stayed as a cover and as the base of autologous cultured epidermis (Lorenti, 1993; Prezzavento *et al.*, 1991). This technology was used in some patients at the Italian Hospital in Buenos Aires until 2007. At present, several laboratories, both state and private, are implementing the culture of keratinocytes (Lorenti, 2012; Centro de Quemados y Cirugía Plástica) as well as chondrocytes (Lucero, 2012) for therapeutic purposes.

The Department of Orthopedics and Traumatology in the Italian Hospital in Buenos Aires was created in 1926, by the initiative of Professor José Valls, and was promoted by the Italian teacher Vittorio Putti from Bologna, one of the forerunners of modern orthopedics. This department is one of the oldest in Argentina and currently works with sections dedicated to the various sub-specialties encompassing the study and treatment of all diseases related to Orthopedics and Traumatology. Since its establishment, successful surgeries have been performed using frozen osseous tissues. In 1944, Dr Carlos E. Ottolenghi, who was also a disciple of Putti, assumed the leadership and gave a great impetus to the scientific department (Hospital Italiano de Buenos Aires). In 1948 Dr Carlos E. Ottolenghi performed the first bone transplant at the Italian Hospital in Buenos Aires from a cadaveric bone piece banked in a frozen state (Muscolo *et al.*, 1987). In 1982 the Center for Research and Studies Orthopedic and Trauma (CINEOT) was created where the bone bank resides (Muscolo *et al.*, 1992).

The Eye Bank of Saint Lucia Ophthalmology Hospital was established in 1984, and it has had uninterrupted operations since 1995. It is the only eye bank enabled and running within the scope of the government of the city of Buenos Aires, as it fulfills its function both in public hospitals of the city and in the private sector. The operators collect eyes from donors and evaluate potential donors and ablations performed in clinics, sanatoriums, funeral houses, judicial morgues and homes (Bouchez and Cittadino, 2006).

In 1994, the Pediatrics Hospital S.A.M.I.C. Professor, Dr Juan P. Garrahan, recognized the need for cryopreserved heart valves for correction of congenital heart defects. After researching the feasibility of reproducing the protocols of Robert Parker (Bank of Homografts, Royal Brompton Hospital in London), who graciously offered it, the first valves stored at 4°C were successfully implanted that same year (nowadays they are stored at −170°C). The hospital opened the Heart Valve Bank in 1996, after having the first legislation to regulate this activity by the INCUCAI.

From that time, SAMIC Professor Dr Juan P. Garrahan at the Pediatrics Hospital has been processing and storing various cardiovascular tissues such as aortic and pulmonary valves, pericardium, ducts, arteries, and veins. They are distributed to private and public centers across the country and even abroad. In 2001, they began to incorporate the production of other tissues, such as amniotic membrane for ophthalmic use and as temporary cover for burn patients in 2002. In 2003, the Bank started skin processing, and in 2007 it produced musculoskeletal system tissue allografts (Hospital Garrahan).

The Blood Products Laboratory at the National University of Córdoba has a tissue bank that procures and processes human bone coming from cadaveric donors for therapeutic use in Dentistry and Traumatology. It was inaugurated in 2003 as a tissue processing center, and in December 2013 it was established as a tissue bank. In this laboratory, bone matrix in powder, fine granules, cubes, sheet (demineralized bone tissue) and block sliced (bicortical bone) are produced. These products are all lyophilized and sterilized by gamma radiation (Laboratorio de Hemoderivados).

At present, the tissue banks authorized by the INCUCAI are summarized in Table 8.1 (INCUCAI, 2014).

Since 2003, in order to promote transplants in Argentina and in countries in the region, the INCUCAI has signed several bilateral cooperation agreements (INCUCAI-Integración regional) with the following countries:

- Spain. A specific collaboration agreement in the field of education and training of human resources in the transplant area was signed in April 2003. Under this agreement signed between both Ministries of Health, the Spanish National Transplant Organization (ONT) has assisted the

Table 8.1. Active tissue banks in Argentina.

	Tissue bank				
Place	**Multi-tissues (Amnion, bone, heart valve, skin)**	**Musculoskeletal**	**Amnion**	**Cornea and sclera**	**Heart valve**
Buenos Aires City	1	4		1	1
Buenos Aires Province		2		1	
Córdoba Province		3		1	
Santa Fe Province			1	1	
Tucumán Province		2		1	
Corrientes Province				1	
Santiago del Estero Province				1	
San Juan Province				1	
La Pampa Province				1	
Total	1	11	1	9	1

INCUCAI in training professionals for addressing family interviews and in the development of the network of hospital coordinators.

- Uruguay. A specific agreement with the National Institute of Donation and Transplantation Cells, Tissues and Organs (INDT) on matters of operational coordination was signed in December 2005. The basis was established for both agencies, and they were empowered to exchange organs and tissues, provided there is no recipient in the country of origin.
- Paraguay. An agreement on cooperation and technical assistance with the Ministry of Health and Social Welfare through the National Institute of Transplantation Ablation (INAT), Paraguay was signed in September 2009.
- Chile. A cooperation agreement with the Ministry of Health of Chile and the Ministry of Health of Argentina for reciprocity procurement and transplantation of organs and tissues was signed in January 2011.
- Peru. An agreement on cooperation and technical assistance with the Ministry of Health of the Republic of Peru through its National

Organization for Donation and Transplantation (ONDT) was signed in August 2012.

- Venezuela. A cooperation agreement with the Venezuelan National Transplant Organization (ONTV) was signed in August 2012.
- Others.

  — The INCUCAI signed an agreement with the State of Bahia, Brazil (November 2011) and the University of Edinburgh, United Kingdom (October 2008) with the aim of developing a program of technical and scientific cooperation.
  — Since June 2003, INCUCAI has integrated the international network called Bone Marrow Donor Worldwide (BMDW).
  — In 2009, INCUCAI signed an agreement with the US National Marrow Donor Program (NMDP), which reduces time and cost for both countries in search of donors and the HPC collections from donors of cord blood units (umbilical and placenta). It also signed an agreement with the European Marrow Donor Information System (EMDIS), for the online computer connection of the Argentine and European Registries.
  — In late 2013, the INCUCAI signed an agreement with the Pan American Health Organization (PAHO) to allow high-resolution HLA typing and low costs; Argentine samples HPC donors.
  — In March 2014, INCUCAI signed a project agreement with the Anthony Nolan Registry, United Kingdom, that will, like the one signed with the United States, reduce time and costs for both countries in search of donors and HPC-donated collections.

Bolivia

Law 1716/1996 on Organ Donation and Transplantation, Cells and Tissues is regulated by the Supreme Decree No. 24671/1997.

This Supreme Decree includes the regulation on practices of donation, ablation, and transplantation of organs, cells, and tissues, which are intended to extend life and rehabilitate people affected by irreversible disease, and the establishment of public health, social, and private security. The organs, tissues, and cells from humans may only be through non-profit donation, and their transplantation in humans carried out only when other treatment options have been exhausted.

The National Secretary of Health through the Direcciones Departamentales de Salud (DIDES) (Departmental Directorates of Health) authorizes the operations of Organs, Cells and Tissue Centers and performs the control, inspection, and supervision of health facilities and the activities related to the practice of ablation and transplantation of organs, tissues, and cells.

The Supreme Decree 1115/2011 regulates Law No. 1716/1996 on Donation and Transplant of Organs, Cells and Tissues, establishing the scope of application; organs, cells, and tissues that can be donated; the donor categories; the recipients; and functions and duties of health facilities and professionals involved in donation and transplantation of organs, cells and tissues (Gaceta Oficial, Bolivia).

In Bolivia, there is a significant demand for corneas, but there is no eye bank. In order to improve care and service, the National Institute of Ophthalmology is expected to create the first eye bank in 2014 (Instituto Nacional de Oftalmología).

Brazil

Since 1997, Federal Law No. 9434 has regulated organ donation (Diário Oficial da União, 1997). It basically states that every individual is a potential multi-organ donor, unless stated against this in the identity card and/or in the driver's license. In 1998, an addendum to the law was adopted, suggesting family consent before organ procurement. The Federal Decree No. 2268 from 1997 established the National Transplant System, where the Ministry of Health is in charge of all procurement and disposition of organs in Brazil, assuring the minimum standards for donor selection and storage of organs. To expedite organ donations and procurement, a National Transplant Organization has been established to supervise regional and local procurement teams, and it also gives to medical institutions the required accreditation to perform organ procurement and transplantation (Herson *et al.*, 2009).

At present, tissue banking activities are regulated by Law No. 28189: General Law on Organ and/or Human Tissues Donation and Transplantation

(March 2004), the Supreme Decree N° 014-2005-SA (May 2005) and Law No. 29471 (June 2010). Resolution 2600N (21/08/2009) approved the Technical Regulation of the National Transplant System that regulates the tissue banks.

The functions of the Sistema Nacional de Transplantes (SNT) (Central Organ of the National Transplant System), shall be exercised by the Ministry of Health through the Coordenação Geral do Sistema Nacional de Transplantes (CGSNT) (General Coordination of the National Transplant System). The State Coordination of the SNT shall be exercised by the Departments of Health of the States and Federal District, or equivalent bodies, through their State or District Coordination Transplantation. The State Secretariats of Health and the Federal District, in accordance with an assessment of its Central Notification of Organ Procurement and Distribution, are authorized to create Procurement Organizations of Organ and Tissue (OPO).

The transplantation of tissues, organs, cells, or body parts can only be performed by established health and specialized teams previously authorized by CGSNT. All tissues, organs, cells, or body parts obtained from deceased donors to its destination should be distributed according to the system of single list.

Tissue banks are defined as health establishments which have physical facilities, equipment, human resources, and appropriate techniques for identification and triage of donors, procurement, processing, storage and distribution of tissues and their derivatives, of human origin, living donors or bodies for therapeutic and research purposes.

In the Regulation the tissue banks are classified in the following categories: Ocular Tissue Bank, Musculoskeletal Tissue Bank, Skin Bank, Cardiovascular Tissue Bank, Umbilical Cord Blood and Placental Bank.

The Brazilian Health Surveillance Agency, Agência Nacional de Vigilância Sanitária (ANVISA) was created by Law 9782/1999. In the federal public administrative structure, the agency is connected to the Ministry of Health. It is a governmental regulatory agency and its mission is to protect and promote public health and to intervene in the risks caused by the production and use of products regulated by health surveillance. This mission must be carried out in coordination with states, municipalities and the Federal District, according to the Brazilian Unified Health

System principles, in order to improve the quality of life of the population. To assure the quality of treatments based on the use of blood, other tissues, cells, and human organs, ANVISA formulates technical rules, inspects services, and monitors the occurrence of adverse reactions caused by technologies available in Brazil. It has created a multi-disciplinary group to produce recommendations for inspection of tissue banks, with 15 multi-disciplinary participants, including the area of radiation sterilization and also recently a technical regulation group on primary packaging features, RDC No. 32, June 2012 (ANVISA).

Table 8.2 shows the number of tissue banks in Brazil.

Table 8.2. Active tissue banks in Brazil.

Type of tissue bank	Number
Eye	41
Skin	2
Heart valve	1
Musculoskeletal	9
Total	53

Chile

Law No. 19451 of 1996 entitled Law on Transplant of Organs and Other Anatomic Parts establishes the transplant and organ donation standards. It is stated that organ and tissue transplants can only be done with a therapeutic purpose at hospitals and clinics that are accredited to fulfill the conditions and requirements carried out in such institutions. The donor should express their consent in a free and informed way, indicating clearly the organs/tissues he or she is willing to donate.

The Law creates an advisory commission within the Health Ministry named National Commission on Organ Transplant that establishes the standards for the organization and functioning of organ and tissue transplants. The 'Sanitary Code' states that cadavers from persons deceased in public or private hospitals or institutions belonging to the Legal Medical Service (Servicio Médico Legal), not claimed within the term established by the Code, could be used for scientific or therapeutic purpose (Ley Chile, 1996).

The law governing Donation and Transplantation of Organs and Tissues is the No. 20673/2013, modification of Article 2 of the Law No. 19451 regarding the determination of who may be considered organ donor. It establishes that any person over eighteen years will be considered as organ donor once dead, unless expressing the contrary when alive (Ley Chile, 2013).

The creation of the National Coordination of Procurement and Transplantation at the Ministry of Health gives the first ministerial and state structure to the donation and transplantation activity.

Tissue banking in Chile started as an idea in 1996. Before 1996 there were only a few small surgical bone banks. The first bone bank was established in 1998, at the Hospital San Borja Arriaran in Santiago, jointly with the Chilean Nuclear Energy Commission. Also in the same year, the first tissue bank was established, the Laboratory of Processing of Biological Tissues Radiation Sterilised (LPTR), at the Nuclear Center La Reina of Chilean Nuclear Energy Commission, with the technical and financial support of the IAEA. Since 2001, this laboratory began to produce tissues for clinical use. It started with the processing of human amniotic membranes and bone, as well as of pig skin, and in 2002 it started with human skin processing (Aguirre and Morales, 2009; Aguirre, 2012; Ogrodnik, 2012).

Colombia

The agency responsible for the surveillance and control of the activity of tissue banking is the National Institute of Food and Drugs, Instituto Nacional de Vigilancia de Medicamentos y Alimentos (INVIMA); it is an adscript to the Ministry of Health Public, conferred by Act 100 of 1993.

Decree 2493 of August 2004 regulates the procurement, donation, preservation, storage, transport, and disposal destination of anatomical components and transplant procedures or implantation in humans.

Currently in Colombia there is an average of 10.7 donors per million inhabitants. In 2007, Bogota reported 17.7 donors per million, the highest figure within the country (Guerrero, December 2012).

Regulations on tissue banking activity are given by the following decrees and resolutions:

- Decree No. 2493/August 2004 regulates the procurement, donation, preservation, storage, transport, and disposal destination of anatomical components and transplant procedures or implantation in humans.

- Resolution No. 5108/2005, standard for the operation of Tissue Banks.
- Decree No. 1011/2006, mandatory System for Quality Assurance in Health Care.
- Decree No. 4295/2007, Technical Quality Standard.
- Decree No. 1164/2002, Manual of management of hospital wastes and similar in Colombia.
- Resolution No. 5108/2005, Good Practice for Tissue Banks.

INVIMA is the organization that protects and promotes the health of the population through risk management associated with the consumption and use of food, drugs, medical devices, and other products subject to sanitary surveillance. It envisages in 2014 being a leading institution nationally and internationally recognized by the National Health Surveillance System, for its transparency, operational effectiveness and technical adequacy, and permanent generator of security and confidence in the population.

The Donation and Transplant Network is formed by a National Coordinator, the National Institute of Health, and five Regional Coordinators in charge of each local authority, with its specific area of influence.

The most active bank is the Bone and Tissues Bank of Cosme and Damian Foundation opened in 1991. Up to 2005 it had processed more than 60,000 grafts, and at least 95% were implanted in a total of 34 hospitals distributed around Colombia and in the five neighboring countries (Cosme y Damian Bank).

There are 14 tissue banks certificated by the INVIMA of Colombia, as shown in Table 8.3.

Among the 14 banks there are only two public banks, the Hospital San Vicente de Paul Bank (bone and skin) in Medellín which provides exclusively for their own patients, and the District Tissue Bank (skin, amniotic membrane and corneas) of Bogotá (Guerrero, November 30, 2012).

Costa Rica

Law No. 7409/1994 authorizes the transplantation of human organs and anatomical materials. Eighteen years after its approval, the Costa Rican law that regulates organ donation has not been in operation. For that reason, in September 2011, File No. 18246 was presented to the

Table 8.3. Active tissue banks in Colombia.

Type of tissue bank	Number
Cornea	6
Skin	2
Amniotic membrane	1
Heart valve	3
Musculoskeletal	2
Total	14

legislature as the Law on Organ Donation and Transplantation and Human Tissues, powered by transplant organizations. Subsequently the Legislature of Costa Rica, in April 2014, approved the Law on Organ Donation and Transplantation and Human Tissues by majority affirmative opinion (Asamblea Legislativa de la repíblica de Costa Rica, 2013). The authorization of the Law for Transplanting Human Organs and Anatomical Materials No. 7409/1994 and its amendments were repealed. From the date of promulgation of this law, the Executive Power will have up to one year for regulating it.

This law regulates the activities related to procurement and clinical use of human organs and tissues, including the donation, extraction, preparation, transportation, distribution, transplantation, and monitoring for therapeutic purposes. The Ministry of Health as the lead health authority is responsible for explicitly authorizing establishments, both public and private facilities, to carry out the process of donation and transplantation of organs and tissues.

The Technical Executive Secretariat Donation and Transplantation of Organs and Tissues will aim at coordinating actions with other agencies of the Ministry of Health, the national institutional coordinator of the Costa Rican Social Security Fund Donation and Transplantation in the private sector, civil society and other related sectors, to ensure transparency, accessibility, timeliness, effectiveness, quality, and safety of the donation and transplantation of organs and tissues.

The active establishments related to tissue banking in Costa Rica are the following (Rojas, December 2012; Rojas, November 2012).

- The Eye Bank of the Social Security, opened in 2001;

- The National Skin Bank, opened in 2010; and cell culture laboratory at Technologic of Costa Rica (TEC), opened in 2011.

Cuba

The legal framework for the donations of organs and tissue in Cuba is established by the Public Health Law from 1983. Its Statutes of the Public Health Law/1988 established the voluntary and free character of donation, its humanitarian character, and that the consent should be informed through a signature and stamp validation of this action. The Ministry of Health, through the National Coordination Office for Transplants, has contributed to the donation culture that exists in the population in general (Morales Pedraza *et al.*, 2009; Otero, 2012). On items 81 and 82 of the Statutes, the Public Health Ministry is instructed to implement mechanisms to accredit those units from the national Health System that are carrying out procurement, processing, and distribution. The regulatory authority is the State Control Center of Medical Equipment that establishes authorization and registers institutions and products (Sánchez Noda, 2000).

The first multi-tissue bank was founded at Frank Pais Orthopaedic Hospital, Havana in 1958. Among the Latin America countries, this tissue bank was the first bank to use radiation, a ^{60}Co source, to sterilize tissues. In 2000, it established an Amniotic Membrane Bank at CEADEN; this bank also uses irradiation to sterilize its products.

In Cuba these are the following tissue banks: bone, cornea, blood vessel, heart valve, skin, and amniotic membrane.

Dominican Republic

In the Dominican Republic the Law Donation and Legacy Organ and Tissue Transplant No. 329 was enacted in 1998. It traces the rules of law governing ablation, transplantation, and available organs, tissues, or anatomical derived materials from humans, for therapeutic purposes, research, and teaching. In it, the remuneration or any compensation or reward for donating an organ or tissue is prohibited. Substantiated in Law 329-98, under the rectorship of the Ministry of Public Health and Social Assistance (SESPAS), the National Transplant Council (CNT) and the National Institute of Transplant Coordination (INCORT) are created, whose functions are the regulation and control of the practice and extraction, ablation, conservation, and exchange of organs and tissue for transplantation, research and education, following the fundamental principles of ethics.

In 1989, Bank Corneas was founded by Dr Luis Cuello Mainardi and Dr Juan Francisco Battle. This bank has worked closely with other banks in Miami, San Diego, and Denver in the United States (Casablanca, 2008), by subscribing to the code of ethics of the Pan American Association of Eye Banks (APABO). Unfortunately, the majority of corneas provided by this bank have been imported (Almánzar, 2013; Santos, 2013).

At present, along the entire Dominican territory, numerous procedures for transplantation of both organs and tissues have been carried out. The Dominican Republic does not have banks for tissue preservation (Almánzar, 2013).

Ecuador

The following information about the legal system for tissue banking in Ecuador was obtained from a manuscript provided by Dr María Candela Ceballos (Ceballos, 2012).

In Ecuador, the first tissue transplantation was an allogenic hand in 1964. It was done in Guayaquil by Dr. Roberto Gilbert Elizalde (Gilbert, 1964a). This hand transplantation was the first in the world, but did not reach the desired success, since the implanted limb was rejected and forced to be subsequently amputated (Gilbert, 1964b).

The first tssue bank in Ecuador was the Eye Bank of the Hospital Abel Gilbert Ponton, better known as Hospital Guayaquil, created in 1981. Its creator, Dr Emigdio Navia, carried out approximately 1,200 corneal transplants until 1996, when the service disappeared after his death.

In 1986, Dr Christopher Cabezas created the first bone bank of Ecuador in the Orthopedic and Traumatology Department of Hospital Carlos Andrade Marín in Quito. In 2002, the musculoskeletal tissue bank at Hospital Alcívar was established by the initiative of Dr Eduardo Alcívar, Orthopedic and Traumatology specialist; it was accredited by the National Regulatory Authority in 2004.

Organ and tissue transplantation and tissue banking activity have not been consistent over the last 20 years as in other countries of the region. This is probably largely due to lack of funding and implementation of policies to promote this activity (Ceballos, 2012).

The new Constitution of the Republic of Ecuador, published on 20 October 2008, in its Article 50 provides that the State shall ensure that any person suffering from catastrophic diseases of high complexity has the right for free specialized care at all levels, and in preferential timely (Registro Oficial No. 449, 2008). The definition of Catastrophic Diseases was established by a Ministerial Decree published in the Official Journal No. 798/2012 (Registro Oficial No. 798, 2011), and which considered catastrophic, rare, and orphan diseases. It includes all types of congenital malformations of the heart and all types of cardiac valves; all types of cancer; brain tumor at any stage and of any kind; chronic renal failure; transplantation of organs: kidney, liver, bone marrow; *sequelae* of severe burns; cerebral arterial venous malformations; Klippel Trenaunay syndrome; Aneurysm thoracic–abdominal; subcomponent of disabilities; external prosthesis of upper and lower limbs; cochlear implants; and orthoses (postural chairs).

Law No. 58/1994, National Law of Transplantation of Organ and Tissues and its regulations, was issued on July 6, 1998 and enacted in the Register Official No. 356 which was updated. The new Law on Donation and Transplantation of Organ, Tissue and Cells was approved by the National Assembly and published in the Official Gazette No. 398 on March 4, 2011 (Registro Oficial No. 398, 2011).

This new law aims to promote transplant activity in the country and consolidate the National Integrated System of Donation and Transplant. Strengthening the public system and acting under the guidance of the State through the National Health Authority, it aims to ensure compliance with the rights established in the Constitution of the Republic, and in strict compliance with the treaty provisions and applicable international instruments on the human genome, genetic data, and bioethics.

This new law guarantees the right to transplantation, through the regulation of activities related to the collection and clinical use of human organs, tissues, and cells, as well as products derived from there, including promotion of donation, extraction, processing, storage, transportation, distribution, and transplantation. A significant change, as compared to the previous law, is that it establishes that dying Ecuadorians as well as foreigners legally resident in the country, in both cases over 18 years old, become donors unless they have expressed in life their willingness to the contrary. They may restrict their affirmative will for donation of certain organs, tissues, and/or cells. The demonstration, restriction, or condition of the will to donate anatomical components will be noted in the identity card in the case of Ecuadorian citizens, and any other identification document in the case of foreigners legally residing in the country.

Another important change in the legislation is the inclusion of regulations for tissue banks, unlike the previous law which only mentioned these in definitions.

The new Law on Donation and Transplantation defines a tissue bank as one unit of technical, specialized and authorized healthcare which obtains, processes, stores, and preserves tissues and/or cells, for subsequent implantation or use for therapeutic and research purposes, under rules to ensure quality from procurement to clinical use.

From 1999 to 2012, Ecuador had a regulator organization for transplantation activity. The National Agency for Organ Transplantation and Tissue (ONTOT), was created by Ministerial Decision No. 1808-A, published in the Official Journal No. 343, as an entity under the authority of the Directorate General of Health, Ministry of Health. Then it was replaced by the National Institute of Donation and Transplantation of Organs, Tissues and Cells (INDOT), created on July 14, 2012. The INDOT is under the Ministry of Public Health and it is responsible for regulation, coordination, monitoring, advocacy, and evaluation of nationwide transplantation activity. The INDOT works with the highest technical standards based on ethical principles of fairness and transparency in strengthening the Integrated National Donation and Transplant. The headquarters of the institution is located in Quito and deployed with zonal coordinators in Pichincha, Guayas, and Azuay (Registro Oficial No 745, 2012). Its mission is to implement public policies on donation and transplantation of organs, tissues, and cells; through regulation, coordination, promotion, provision, control, monitoring, and evaluation of the activity of donation, extraction, preservation, allocation, and transplantation of organs, tissues, and cells, strengthening the Integrated National Donation and Transplant to the highest technical standards under the bioethical principles of respect, fairness, and transparency.

At present Ecuador has accredited the following tissue Banks (INDOT):

- National Tissue and Cells Bank (BANTEC), which has three subsidiaries, the largest located in Quito, Cuenca, and Guayaquil;
- Tissue Bank, Luis Vernaza General Hospital, Guayaquil;
- National Oncologic Institute "Dr Juan Tanca Marengo" (SOLCA), Guayaquil.

El Salvador

The Legislative Assembly of the Republic of El Salvador promulgated Decree No. 955/1988, the Health Code. The Minister of Public Health and Social Assistance decreed the Law of Organ and Tissue Transplant, General Dispositions and Decree No. 291, a reform to the Health Code, where it amended the name of Section 19, Chapter II of Title II of the Health Code as follows: "section nineteen organ or tissue transplant." It reformed Article 128 as follows: The Ministry of Health and Social Assistance is the governing body of the national policy on human organ and tissue for medical and scientific purposes, developed in consultation with the National Transplant Council. It was added to Section 19, after Section 128, the following dispositions: Creation of the National Transplant Council as a consultative and advisory body to the national transplant policy, which will be regulated by the Minister of Health and Social Assistance or his substitute. The practice of transplantation of human organ or tissue will be performed in strict accordance with the rules of ethics and based on principles of equity, justice, solidarity, voluntarism, and without distinction of any kind.

This Code applies to organ and tissue banks that hold deposits of materials or tissues for future use by individuals or research, and to the therapeutic use of human organs or tissues to replace a diseased organ or tissue, or function otherwise healthy.

El Salvador does not yet have a tissue bank (Primer medio digital salvadoreño, 2014).

Guatemala

The decree No. 91/1996 on Law enables the Provision of Human Tissues and Organs. The Ministry of Public Health and Social Assistance controls the application of this law and its regulations as well as programming, coordination, monitoring, and evaluation of the activities regulated in it.

In 2006, this decree was regulated by the Ministry of Public Health and Social Assistance for Law Enforcement for the provision of human organs and tissues. Afterwards this decree was repealed in 2010, and it approved the Law for the Provision and Transplantation of Organs and Human Tissue.

The Ministry of Health and Social Assistance released the Government Agreement 440-2013, which contains regulations for the establishment, operation, and control of banks of cornea and sclera, public or private.

Honduras

The Decree Law on Transplant and Removal of Organ and Tissues was promulgated by Decree No. 131/1982. It was recently abrogated by a new one that regulates the Law on Donation and Transplantation of Anatomical Organs in Humans, Decree No. 329-2013.

The new law contains rules for transplantation or disposition of organs, tissues, or anatomical materials derived from humans for therapeutic purposes, research, and teaching; also included are ova, semen, embryos, cells, or any other tissue or body material that are implanted by clinical or laboratory means, therapeutic, genetic, or aesthetic. It adds that donation is a voluntary and free act. Biogenetic experimental or clinical activity concluding in the conception or birth of humans with genetic cloning is prohibited (La Gaceta, 2014).

Mexico

The following information about the legal system for tissue banking in Mexico is obtained from a manuscript provided by Dr María Esther Martínez (Martínez Pardo, 2013).

At present, the legal aspects concerning the donation and transplantation of organs, tissues, and cells derive from the fourth article of the Political Constitution of the Mexican United States, which states the right of all persons to the protection of health. On February 7, 1984, the General Health Law, Ley General de Salud (LGS) was published in the Federal Official Newspaper (DOF) and this Law, in its 14th title of Donation, Transplants and Loss of life, has been updated in 2009 and 2011. The 14th title of the LGS and Regulation for Sanitary Control, published in February 1985, and its amendment in 2012 describe the procedure for the disposal of organs, tissues, and human cadavers (Ley General de Salud, 2012).

Due to modifications to the LGS, the Health Secretariat has two agencies for regulation and control of activities of donation and transplantation of organs, tissues, and cells: The National Transplant Center, Centro Nacional de Trasplantes (CENATRA) and the Federal Commission for the Protection against Sanitary Risk, Comisión Federal para la Protección contra Riesgo Sanitario (COFEPRIS).

CENATRA, founded in 2000, is the agency of the Health Secretariat responsible for the stewardship of the National Transplant System. It has as its primary mission to organize and encourage the donation and transplant programs in health institutions. Its functions include the issuance of policies and monitoring the distribution and allocation of organs and tissues donated by the people. In other words, CENATRA looks out for organ/tissue procurement up to transplant on the recipient; this agency also keeps the registers of donor–recipient to have the traceability of organ/tissues at national level. CENATRA is organized in the following directions: General; National Transplant Register; National Planning, Teaching and Coordination. The sub-directions are: Norm and Legal Affairs; Administration (Centro Nacional de Trasplantes).

In 2001, the Comisión Federal para la Protección contra Riesgos Sanitarios (COFEPRIS) was created. This is a decentralized organization of the Health Secretariat with technical, administrative, and operational autonomy, whose mission is to protect the population against sanitary risks. Since April 13, 2004, the COFEPRIS is in charge of the sanitary control of the processing, disposal, and transplantation of human cells, organs, tissues, and their components. COFEPRIS has also granted the sanitary license to the tissue banks in Mexico and to the tissue procurement establishments, since 2004 (COFEPRIS).

In general, the tissue banks established in Mexico are the five specialty banks that use gamma radiation as a tissue sterilizing agent, and these banks have implemented Good Manufacturing Practice (GMP), Good Laboratory Practice (GLP), and Good Tissue Practice (GTP). These banks have the documentation required by a quality management system (QMS) but only two banks have the ISO 9001: 2008 certification of their QMS.

At present, the health regulatory organizations recommend following Mexican Official Norm 241, GMP, for establishments to manufacture medical devices (Norma 241 SSA, 2012) for all processes, from procurement/handling up to the clinical application of blood, bone marrow, organ, tissue, and cells.

According to CENATRA records, Mexico has 69 tissue banks, as shown in Table 8.4, but for a country with more than 119 million inhabitants, a particular concern of CENATRA is the future growing number of tissue donations when tissue banks are not growing at the same speed; there is not sufficient number of tissue banks in the country (Martínez Pardo *et al.*, 2009).

The Mexican tissue banks that use gamma radiation to sterilize tissues and conditioning materials are the following:

- Radiation Sterilized Tissues Bank, Banco de Tejidos Radioesterilizados (BTR), at the National Institute of Nuclear Research, Instituto Nacional de Investigaciones Nucleares (ININ). This bank produces human and

Table 8.4. Active tissue banks in Mexico.

Tissue	Number of banks
Cornea	44
Skin	6
Amniotic membrane	2
Bone and musculoskeletal	5
Cardiac valves	1
Stem cells	4
Germinal cells	6
Total	68

pig skin and amnion, and has a quality management system in force, certified under ISO 9001:2000 since August 1, 2003 (Martínez Pardo and Magaña, 2007).

- Bone and Tissues Bank, University Hospital "Dr. José E. González", Banco de Huesos y Tejidos (BHT), at the Autonomous University of Nuevo Leon State, Monterrey City.
- Novoinjertos S.C., located in Mexico City; this bank produces bone tissues and has the ISO 9001:2008 as well as the ISO 13485 certifications.
- Tissue Bank of the Mexican Biotechnology Institute, Banco de Tejidos del Instituto Mexicano de Biotecnología (BT-IMB); this bank is located in Zapopan City, Jalisco State, and produces bone and musculoskeletal tissue.
- Mexico State Tissue Bank, Banco de Tejidos del Estado de Mexico (BTEM); this bank produces skin, and is located in Toluca City, Mexico State.

All these banks sterilize their products by using the ^{60}Co industrial gamma irradiator that belongs to the Gamma Irradiator Department of ININ.

International relations have been established with the following organizations:

- Red Consejo Iberoamericano de Donación y Trasplante (Iberoamerican Council Net of Donation and Transplant)
- Sociedad de Trasplante de América Latina y el Caribe (Latin America and the Caribbean Transplant Society)
- Spanish Organización Nacional de Trasplantes (National Trasplant Organization)
- Grupo Punta Cana
- European Transplant Coordinators Organization

Nicaragua

Chapter V of the General Health Law No. 423/2002 refers to the donation and transplantation of organs and tissues, and later the Penalty Law No.

641/2008, in Article 346, mentions trafficking and removal of human organs and tissues.

Recently, the National Assembly of Nicaragua approved Law No. 847 on Donation and Transplantation of organs, tissues, and cells, which was approved on October 9, 2013 (La Gaceta, 2013). This law aims to establish the rules governing the donation and procurement of organs, tissues, and cells of living human individuals or bodies to be transplanted for therapeutic purposes, teaching, and research in humans. Also subject to this law is the autologous transplantation of embryonic stem cells, eggs and sperm, and blood and its components. This law is of public order and social interest, and the Ministry of Health is the governing body for the implementation of the Act, its Regulations and the provisions issued to that effect.

This law covers all aspects of the regulation and control of the activity of donation and transplantation of organs, tissues, and cells of living individuals or human cadavers in the following sections: General Provisions; Procurement of tissues and organs from live donor cells related to persons and dead; Exceptions, priorities, rights and prohibitions of children and adolescents; Creating list of recipients of the National Registration System transplants of organs, tissues and cells; Registration and accreditation of professionals; Accreditation, certification and approval of establishments; Creation of the Nicaraguan Transplant Organization (ONITRA) which is established as the operational technical unit; Requirements for carrying out the transplantation of organs, tissues and cells; Tissue banks (establishment, operation, registration and issue of materials available — organs, tissues and cells); Rights, duties and guarantees of person donors, receivers and accompanying family members; Promotion and publicity, education and training; Administrative proceedings to punish infringements of this Law; Reform Act No. 641 of the Penal Code.

This law became effective from April 2014.

Panamá

On February 2010, the National Assembly approved Law No. 3, General Law on Transplantation Anatomical Components, which created the Panamanian Transplant Organization (OPT).

Among the first actions of this organization was the creation and updating of the National Transplant Registry, and also the National Waiting List where all people waiting for a transplant of an anatomical component are listed.

In 1985, the first eye bank of the Club de Leones of Panamá (Banco de ojos) was created.

Paraguay

Law No. 1246 on Human Organ and Tissue Transplant was enacted in 1998 (Ley 1246, 1998). The organ and tissues ablation of people with confirmed brain death and of living persons, for the purpose of transplantation into other living humans, is governed by the provisions of this law, except for tissue and naturally renewable materials separable from the human body. Ablation and implantation of organs and tissues may be performed when other available means of science have been exhausted as a therapeutic alternative for the health of a given patient. These practices are considered current technique and not experimental art.

Under this law, medical procedures referred for transplantation may only be made by doctors or medical teams registered and enabled for this purpose by the supervisory body, which is the National Institute of Ablation and Transplantation, Instituto Nacional de Ablación y Trasplantes (INAT). This will require, in all cases, sufficient medical training and experience in the specialty. Authorization to heads and deputy heads of teams and professionals will be awarded by the INAT, informing of the management to the Ministry.

In Asunción, the Fernando Oca del Valle Eye Bank Foundation has been doing corneal transplants since 1978, with marked success. This eye bank was named after this person because he was the first cornea donor of Paraguay (Fundación Banco de Ojos).

The first multi-tissue bank in Paraguay is being set up by the Social Security Institute (IPS), which will process, store and distribute materials.

Peru

Peru has General Law No. 28189 on Donation and Transplantation Organ and/or Human Tissue, enacted in March 2004 (Ley No. 28198, 2004). It is regulated by Supreme Decree No. 014-2005-SA (May 2005) and the following associated regulations:

- Ministerial Resolution approving 999-2007/MINSA Health Technical Standard for Accreditation of Health Donors Facilities — Transplanters;
- Law No. 29471/Dec 2009 that promotes procurement, donation, and transplantation of organs or human tissues; this law is regulated by Supreme Decree No. 011-2010-SA;
- Administrative Directive for implementing committees in establishments — transplant donor health and transplants;
- Administrative Directive No. 169, MINSA/ONDT-V0.1, issued in December 2010;
- Technical Guide for the performance of procedures for the diagnosis of brain death, issued in March 2011;
- Technical Guide for the selection of cornea donors, issued in May 2011.

The tissue banks have implemented Good Manufacturing Practices (GMP), issued by the Directorate General of Pharmaceuticals and Drugs, Dirección General de Medicamentos y Drogas (DIGEMID) of Peru, and is applying the Guide for Tissue Banks Operation, Guía para la Operación de Bancos de Tejidos, IAEA Document RLA6/062 2012.

The National Organization for Donation and Transplantation, Organización Nacional de Donación y Trasplantes (ONDT) is the technical body that regulates transplantation activity of organs, tissues, and cells.

Peru has established the following four tissue banks:

- A multi-tissue bank located at the National Institute of Child Health, Instituto Nacional de Salud del Niño (INSN), the Tissue Bank Rosa Guerzoni Chambergo, established in 1996, where amniotic membrane, bone, and skin are produced;
- An eye bank at the National Institute of Ophthalmology, established in 1970;
- A tissue bank at the Loayza Archbishop National Hospital, opened in 2011.
- The Organ Bank and Histocompatibility Laboratory at Rebagliati National Hospital of Social Security of Health, Seguro Social de Salud (EsSalud).

The limited number of donors and the requirement of materials for grafting, led staff of the Tissue Bank and Laboratory of Radiation Medical Products (LIPM) of IPEN to plan country projects on tissue engineering and scaffolds production to produce new tissues (Castro Gamero and Herrera, 2013).

Uruguay

The following information about the legal system for tissue banking in Uruguay was provided by Dr Inés Álvarez and Dr María del Carmen Saldías (Alvarez and Saldías, 2013).

The regulation of transplantation activity began in 1971 with Law 14.005, and was the first legal framework in Latin America. This legal instrument implemented a unique national and governmental organization operating since 2003 only within an "opting-in" consent system, and a national donor registry of pre-stated acceptance/refusal consents. The first model was based on the altruistic and voluntary decision of any person during his or her life or relative if the deceased person had no voluntary consent. The consent has a testamentary value in favor of every person from the population. So, referred to this legal criterion, the deceased donor belongs to the whole society.

The unique and public organization called National Institute of Donation and Transplant of Cells, Tissues and Organs, Instituto Nacional de Donación y Trasplante de Células, Tejidos y Órganos (INDT) by Decree 248/2005, previously called Banco Nacional de Órganos y Tejidos (BNOT), is responsible for the regulations, policies, and management of donation and transplantation. It was created and regulated in 1977, and opened its doors in 1978 (Decree of Creation, BNOT, No. 86/1977). The organization has a proper budget from the public financial law (Law 16.343 1993 and amendments). The Fund of National Resources (FNR) is the financial support for clinical transplantation.

Uruguay has maintained continuous governmental policy in donation and transplantation, so over the last decade many modifications were implemented in the legal framework in order to update the necessities of clinical transplantation. In this way, a strategic plan by promoting a new legal framework according to scientific progress in clinical transplantation was developed by laws and decrees regulating encephalic death, presumed donation, and security of cells and tissues, as well as the creation of the registration of non-related donors for hematopoietic stem cells, and the re-engineering of tissue banking.

The Transplant Law 17668/03 includes the following: (1) the special presumed donors concept (opting-out criteria), for adult persons who did not refuse donation and when the cause of death requires a coroner intervention; (2) Consanguineous living donors and non-related donors only for Hematopoietic Stem Cells (HSC) transplantation; and (3) xenotransplantation practice.

Through the 287/2004 Government Resolution, Sistema Nacional de Registro, Tipificación y Búsqueda de Donantes de Médula Ósea y Progenitores de Otros origenes (SINDOME) was instituted; this is a national register related to HSC. Implementation, organization, management, and development of this national register were assigned to the Immunogenetic and Histocompatibility Laboratory of the INDT.

By the 160/2006 Governmental Decree, the legal normative for tissue and cell security and traceability transplantation were completed.

Recently, by the 18362/2008 Law, INDT was designed as a Decentralized Organism of the Secretary of Health, National Health Ministry. The Article 337 sets up the National Stem Cell and Cord Bank for Public Use Program.

The last Transplant Law 18968/2012 was extended, to respond to the clinical needs related to the waiting list of recipients for transplantation and the presumed donors concept (opting-out criteria) applicable to all adult deceased persons, who did not refuse donation in his or her life. This new legal scope will provide an important improvement in cell, tissue, and organ donation, to fulfill patients' clinical needs.

Decree 272/013 approved the work restructuring of INDT, establishing the National Bank and Public Umbilical Cord Blood, and the Laboratory of Cell and Tissue Engineering in the field of the INDT-MSP.

Within governmental and institutional relations, INDT depends on the Public Health Secretary, and is in charge of the Legal Normative fulfillment and administrative responsibilities related to transplantation policies.

From an academic point of view, INDT has been operating within the State University Medical School (UDELAR) as a specialized area, with three postgraduate specializations: Transplant Coordination, Tissue Bank for Transplant, and Inmunogenetics for Transplant.

INDT is actually organized by the following (1) Statistics and Register of Illness — Donors, and Donor Refusal; (2) Donors Procurement Unit, responsible for donor promotion and procurement and maintenance of deceased donors; (3) Histocompatibility and Immunogenic Laboratory, responsible for maintenance of the unique waiting list for transplant organs and the assignment organ unit; (4) National Multi-Tissue Bank (NMTB), responsible for the retrieval, processing, validation, and distribution of tissues for therapeutic purposes (tissue production of NMTB includes cornea, cryopreserved vessels, gamma radio-sterilized skin, bone, sclera, fascia lata, and amnion); (5) Cellular products responsible for Bone Marrow Transplantation with HSC Donor Register and National and Public Bank of Umbilical Cord; (6) Consortium of Tissue Bioengineering with the Laboratory of Cells and Tissue Engineering.

Uruguay has integrated in its tissue banking programs and technological development the assistance provided by the IAEA since 2001, contemporaneously with the update of its legal framework, in order to upgrade and optimize tissue procurement procedures to ensure high quality and safety.

The contribution provided by the IAEA programs allowed many technical improvements according to international standards on tissue banking; training of human resources; development and implementation of Tissue Quality System and Risk Management; implementation of radiation sterilization technique, initially carried out in the Co-60 Irradiation facility of the National Atomic Energy Commission of Argentina and at present carried out at INDT, using a Gammacell 220-Nordion, which was provided by the IAEA through the National Cooperation URU/7/005; development of multi-center international scientific research programs, through IAEA Coordinated Research Projects (CRP).

In the fields of tissue bioengineering and cell therapy, a new developing activity is going on in a new physical area, with the purpose of developing and obtaining the best biomaterials for therapeutic application, to apply in all the integrated health systems of Uruguay.

Venezuela

The following information about the legal system for tissue banking in Venezuela was provided by Dr Elizabeth Merentes (Merentes, 2013).

The public health system of the Bolivarian Republic of Venezuela depends completely on the Ministry of the People's Power for Health (MPPS) (Milanés, 2012). Up until now there has been no specific legislation that regulates the procurement and clinical use of tissues of human origin in line with quality and safety standards. Nevertheless, there is a new law governing the Donation and Transplant of Organs, Tissues and Cells in Humans, passed on November 25, 2011 and published in the Official Gazette Nº 39.808 (Gaceta Oficial, 2011). This law replaces the December 1992 law for the Transplantation of Organs and Anatomical Materials in Humans. The aim of the new law is to regulate the procedures involved in the donation and transplantation of organs, tissues, and cells in humans for research, teaching, or clinical purposes, throughout the national territory, based on the right to health as stipulated in the constitution, laws, treaties, agreements, and international conventions signed and ratified by the Bolivarian Republic of Venezuela.

The Venezuelan National Transplant Organization, Organización Nacional de Trasplante de Venezuela (ONTV), established in 1997, is the sole entity authorized by the Ministry of the People's Power for Health to coordinate all matters related to the donation and transplantation of organs and tissues (Informe Organización Nacional de Trasplante). In 2003, the Ministry of Health launched the System for the Procurement of Organs and Tissues (SPOT). SPOT is a network of interdependent and interacting public and private institutions, with qualified staff that work together in a harmonious, systematic, and sustained way for the detection, procurement, maintenance, allocation, and transport of human organs, tissues, and cells from cadaver donors for transplant purposes, in order to effectively satisfy the demand of waiting patients throughout the national territory. Health institutions, authorized transplant centers, hospital transplant coordinators, and the Institute of Immunology are all involved in this process.

In 2006, the MPPS, in order to guarantee and uphold the constitutional principle of the right to life and health, launched the National Program for the Donation and Transplant of Organs, Tissues and Cells

(PNDTOTC) (Milanés, 2012), separating it from the Kidney Health Program. This is a technical-administrative unit, whose function is to ensure the availability of transplants throughout the nation and thus improve the quality of life of the patients that require them. It is attached to the General Directive for Health Programs of the Office of the Vice-Minister for Collective Health Networks, which coordinates the formulation, implementation, monitoring and evaluation of policies and strategies for the various procedures related to the donation and transplant of organs, tissues, and cells in humans for clinical, research, or teaching purposes in order to provide a timely, efficient, effective, and high-quality response to all those that require a transplant of any kind.

Organizations involved in the donation and transplantation of organs, tissues, and cells is schematized in Figure 8.3.

It is important to highlight that in Venezuela presumed consent is established in Article 27 of the current law (ONT, 2006), which states that "Every legally competent person of legal age, who has been diagnosed as dead, shall be presumed to be a donor of organs, tissues and cells for clinical purposes, unless he or she makes a statement to the contrary." Those persons who do not wish to donate all or any of their organs, tissues, and cells can submit a statement to this effect to the National System of Information regarding the Donation and Transplant of Organs, Tissues, and Cells of the MPPS, as MPPS retains the instruments and mechanisms for this purpose, as provided by the law and set out in Article 15 which states that "The National System for Information on Donation and Transplant guarantees the availability of the instruments and mechanisms necessary for a person to express their will against donating some or all of their organs, tissues or cells." The desire to not be a donor should be expressed via the National System of Information for the Donation and Transplant of Human Organs, Tissues and Cells (SINIDOT), which came into operation in November 2012.

Up until now in Venezuela only a few tissue banks have been created within the public health system (Merentes, 2012). The first bank to be established was an eye bank in 2005: the Venezuelan Bank of Ocular Tissues (CONBATOV), located in the Ophthalmology Service at the Francisco Antonio Rísquez Hospital. Currently, the Coordination of the Central Eye Bank operates from the Dr Jesus Yerena de Lidice Hospital,

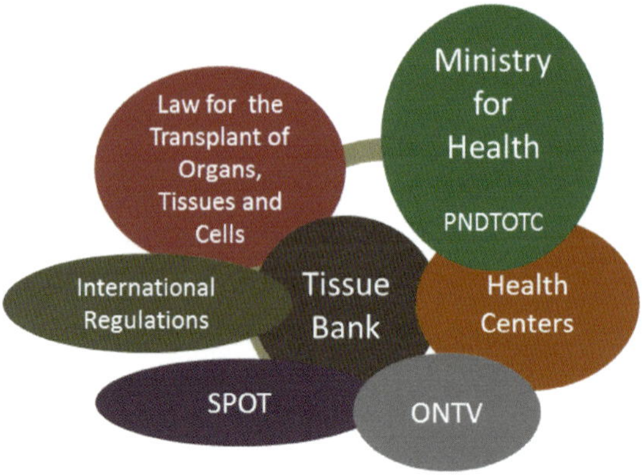

Figure 8.3. Legal regulations and institutes involved in matters pertaining to tissue banks in Venezuela.

in Caracas. This institution is a branch of the MPPS and is attached to the PNDTOTC where it operates together with SPOT. The central aim of this bank is to obtain, analyze, store, document, and distribute corneas obtained from deceased donors. In 2012, another eye bank was set up at the San Cristobal Central Hospital in Táchira State. The main purpose of these banks is to guarantee the timely, free, safe, and universal availability of organs, tissues, and anatomical eye materials to the Venezuelan public, either at its headquarters or affiliated institutions. Corneal tissue is selected according to the standards of quality of the Eye Bank Association of America (EBAA) and the Pan American Association of Eye Banks (APABO) as well as the ONTV and the SPOT system at a national level.

In addition, the Cardiovascular Bank at the Dr Gilberto Rodríguez Ochoa Latin American Children's Cardiology Hospital was opened on January 6, 2009. In this bank, cardiovascular homografts are processed, preserved, and stored for their clinical application in the reconstructive surgery of congenital heart and large blood vessel malformations. The estimated initial storage capacity is of 140 cardiovascular tissues including valves, pericardial, and vascular segments, but this could be increased depending on demand, thus ensuring their free, equitable, and immediate availability to patients in need. Staff training was carried out at the National Bank of Organs and Tissues in Uruguay and the tissue bank in La Coruña, Spain.

The Ministry of the People's Power for Health, according to Resolutions 166 and 167 published in the Official Gazette No. 39210 on June 30, 2009, has authorized the Children's Cardiology Hospital to develop its own system for the procurement and transplant of hearts for children and adolescents with heart diseases. Similarly, the Cardiovascular Tissue Bank has been authorized to create a professional committee to oversee the System of Procurement and Heart Transplants. This committee will be in charge of organizing the order in which the transplant of hearts or cardiovascular tissues in pediatric patients treated in the hospital takes place. In the same Official Gazette, No. 167, the Ministry published a resolution for the creation of a national registry of cardiopathic children and adolescents awaiting surgery.

Recently on May 23, 2013, the Skin Tissue Bank at the Coromoto Hospital in Maracaibo, Zulia State, was inaugurated. This project was started by Dr Tulio Chacín seven years ago and was finally brought to fruition thanks to the support and assistance of the Government of the Republic of Argentina and funding by the Venezuelan National Petroleum Company (PDVSA). Skin is irradiated at the industrial sterilization plant, by gamma rays from a Cobalt-60 source. The center, PEGAMMA, is located at the Venezuelan Institute for Scientific Investigation (IVIC) attached to the Ministry of the People's Power for Science and Technology and Intermediate Industries (MCTI).

Related with prospects of new technologies in the tissue banks in Venezuela, it has recently been proposed that tissue bioengineering should be a permanent component of the structure of a tissue bank. Venezuela should implement new tissue bioengineering technologies, in order to address the shortage of donor material in the medium term. These biotechnologies are already being developed in the country in several research institutions and universities (Venezuelan Institute for Scientific Investigation — IVIC, Central University of Venezuela — UCV, Simon Bolivar University — USB) but further impetus is required in order to apply these new technologies at a clinical level. It is thus important to promote the transfer of technology with countries where the technology is more advanced, taking advantage of the partnerships that exist between regional countries such as Spain, Brazil, and others (Merentes, 2012).

The development of this field could contribute to the creation, development, and consolidation of tissue banks in Venezuela. Specifically, *in vitro* skin and cornea production units should be created, as the current donation rate is 4.51 donors per million inhabitants (ONT-Venezuela, 2013), which is low compared to other Latin American countries. One of the difficulties to be overcome is the lack of information about the importance of donating organs and tissues. In addition, very little about the current situation of the tissue banks in Venezuela is known, as this is an area that has only recently been developed in this country, in spite of its great importance for public health. High-impact outreach campaigns should thus be undertaken at all levels, and networks for the establishment of interrelationships between the national tissue banks and the other institutions involved — hospitals, ministries of health, and universities — should be developed (Merentes, 2012).

CONCLUSION

The regulatory system development related to donation and transplantation of tissues and cells varies significantly from country to country in Latin America. On one hand, the legislation of some countries is complete, updated, and recent. On the other, there is a lack of regulations specific for this purpose, which are different from the decrees and directions reflected in general laws of health regulation. Recently there has been an attempt to reduce the differences between the various countries through governmental efforts, regional agreements, establishment of the Network Council of Donation and Transplant (Red Consejo Iberoamericano de Donación y Trasplantes, RCIDT), ARCAL (Acuerdo Regional de Cooperación para la Promoción de la Ciencia y la Tecnología Nucleares en América Latina y el Caribe) — IAEA projects, and bilateral agreements between countries of Latin America. Thus, there has been progress in the harmonization of the laws in the various countries, human resources training, and strengthening of the activities of cell and tissue banking in Latin America.

ACKNOWLEDGEMENTS

To colleagues María Esther Martínez Pardo (Mexico), María del Carmen Saldías and Inés Alvarez (Uruguay), María Candela Ceballos (Ecuador), and Elizabeth Merentes (Venezuela) for collaborating with the manuscripts related to the legislative situation and the tissue banking activity in their respective countries. To Liliana Bisignano (Argentina), Alicia Lorenti (Argentina), Monica Mathor (Brazil), Linda Guerrero (Colombia), Emma Castro Gamero and Renee Herrera (Peru), and Miguel Rojas Chavez (Costa Rica) for the information provided related to the legislation and the tissue banking activity in their respective countries.

REFERENCES

Aguirre Herrera P, Morales Pedraza J (2009) The impact of the International Atomic Energy Agency (IAEA) program on radiation and tissue banking in Chile, *Cell and Tissue Banking* **10**(2), 153–156.

Aguirre Herrera P (2012) Final Report Project ARCAL CVIII–RLA 6026.

Almánzar Núñez CJ (2013) Tesina: Propuesta modelo banco de tejidos osteotendinosos, inspirado en el modelo español, para el hospital traumatológico Dr. Ney Arias lora, Santo Domingo, República Dominicana, Organización Nacional de Trasplantes, España, Master Alianza en Coordinación de Trasplantes y Establecimientos de Tejidos.

Álvarez I, Saldías MC (2013) Manuscript provided to the author: Material para Capítulo 11 sobre: Legal system for tissue banking in Latin America, Uruguay.

Álvarez I (Compilator) (2014) TCR Editorial, Costa Rica. Guía para la Operación de Bancos de Tejidos.

Alvarez I, Morales Pedraza J, Saldías MC, Pérez Campos H, Wodowóz O, Acosta M, Vicentino W, Silva W, Rodríguez G, Machín D, Alvarez O (2009) The impact of the International Atomic Energy Agency (IAEA) program on radiation and tissue banking in Uruguay: Development of tissues quality control and quality management system in the National Multi-Tissue Bank of Uruguay, *Cell and Tissue Banking* **10**(2), 173–181.

ANVISA-Brazil, http//portal.anvisa.gov.br.

Banco de ojos del Club de Leones de Panamá, http://www.clubdeleonesdepanama.org/ObrasBanco.html.

Bolgiani A (1991) Banco de Piel — Primera parte, *Revista Argentina de Quemaduras* **6**(1–2), 20–21.

Bouchez MI, Cittadino E (2006) Curso Superior de Auditoría Médica, Asociación de Médicos Municipales, Buenos Aires. Monografía: Análisis de costos de funcionamiento del Banco de Ojos Hospital Oftalmológico Santa Lucía.

Casablanca FJ (28 April 2008) Listin Diario, Puntos de vista, "El problema de la donación de córneas.

Castellanos CJ (1985) Panorama histórico de los trasplantes en México, *Cirugía y Cirujanos* **53**(5), 195–202.

Castro Gamero E, Herrera R (December 2013) Personal communication to the author.

Castro Gamero E, Morales Pedraza J (2009) The impact of the International Atomic Energy Agency (IAEA) program on radiation and tissue banking in Peru, *Cell and Tissue Banking* **10**(2), 167–171.

Ceballos MC (2012) Manuscript sent to the author: Legislación y Bancos de Tejidos en el Ecuador.

Centro de Quemados y Cirugía Plástica, Sanatorio Británico, Rosario, Santa Fe, Argentina, http://www.quemaduras.com.ar/cultivo_de_piel.html.

Centro Nacional de Trasplantes, México, www.cenatra.salud.gob.mx.

COFEPRIS (Comisión Federal para la Protección contra Riesgos Sanitarios), México, www.cofepris.gob.mx.

Cosme y Damian Bank, Colombia, http://www.cydbank.org/historia.php.

Declaración de Mar del Plata, (2005) Cumbre Iberoamericana, Red/ Consejo Iberoamericano de Donación y Trasplante.

Diário Oficial da União, Brasília, 5 de Fevereiro de 1997, Seção 1, Lei Federal no 9.434.

Fundación Banco de Ojos Fernando Oca del Valle, Asunción, Paraguay, http://www.bancodeojos.org.py.

Gaceta Oficial, Bolivia, http://www.gacetaoficialdebolivia.gob.bo/normas/view/139720.

Gaceta Oficial No. 39.808, República Bolivariana de Venezuela, 25 Noviembre 2011.

Gilbert R (1964a) Transplant is successful with a cadaver forearm, *Med Trib Med News* **5**, 20–22.

Gilbert R (1964b) Hand transplanted from cadaver is re-amputee, *Med Trib Med News* **5**, 23–25.

Guerrero L (November 30, 2012) Forum: La tecnología de irradiación y los bancos de tejidos: retos y perspectivas, Mexico D.F. Presentation: Bancos de tejidos en Colombia.

Guerrero L (December 2012) Final Report Project ARCAL CVIII — RLA 6026Asamblea Legislativa de la República de Costa Rica, Expediente No. 18.246, Ley de donación y trasplante de órganos y tejidos humanos, October 1, 2013.

Herson MR, Mathor MB, Morales Pedraza J (2009) The impact of the International Atomic Energy Agency (IAEA) program on radiation and tissue banking in Brazil, *Cell and Tissue Banking* **10**(2), 143–147.

Hospital Garrahan, Buenos Aires, Argentina, http://www.garrahan.gov.ar/index.php/equipo-de-salud/banco-de-tejidos-bthg?showall=&start=1.

Hospital Italiano, Buenos Aires, Argentina, Historia, http://www.hospitalitaliano.org.ar/ortopedia/index.php?contenido=ver_seccion.php&id_seccion=7676.

IAEA (2007) Radiation Sterilization of Tissue Allografts: Requirements for Validation and Routine Control. A Code of Practice.

Instituto Nacional de Oftalmología de Bolivia, http://www.ino.gob.bo/index.php/campana/infrae/bancos.

INCUCAI-Historia, https://www.incucai.gov.ar/index.php/comunidad/historia-del-trasplante/avances.

INCUCAI (May 2014) Data from Dirección Científico Técnica.

INCUCAI, http://www.incucai.gov.ar.

INCUCAI — Integración regional, http://web3.incucai.gov.ar/index.php/institucional/.

INCUCAI — SINTRA, http://sintra.incucai.gov.ar/.

INDOT, http://www.donaciontrasplante.gob.ec/indot/?page_id=525.

Instituto Nacional de Oftalmología, http://www.noticiasfides.net/g/sociedad/instituto-nacional-de-oftalmologia-preve-crear-el-primer-banco-de-ojos-para-el-2014--19867/.

ISO 11137:1995, Sterilization of Health Care products — Radiation.

ISO 11137:2006, Parts 1, 2 and 3 Sterilization of Health Care Products — Radiation.

Kairiyama E (Compilator) (2014) TCR Editorial, Costa Rica. Código de Prácticas para la esterilización de Tejidos para Injerto con Radiación Ionizante: Validación y Control de Rutina.

Kairiyama E, Morales Pedraza J (2009) The impact of the International Atomic Energy Agency (IAEA) program on radiation and tissue banking in Argentina, *Cell and Tissue Banking* **10**(2), 139–142.

Kairiyama E, Bolgiani A, Lorenti A, Benaim F (1994) Oral presentation: Skin allografts sterilized by gamma radiation, *Proceeding of Third European Conference of EATB on Tissue Baking*, VIC, IAEA, Vienna, Austria.

La Gaceta, Diario Oficial de la República de Honduras (26/04/2014), NUM. 33.412, http://www.lagaceta.hn/.

La Gaceta No. 207 (31/10/2013), Nicaragua, ley No. 847, http://legislacion.asamblea.gob.ni/.

Laboratorio de Hemoderivados, Universidad Nacional de Córdoba, http://unc-hemoderivados.com.ar/prodTejidos.php.

Ley 1246 (1998), Trasplante de Órganos y Tejidos Anatómicos Humanos, Paraguay, http://paraguay.justia.com/nacionales/leyes/ley-1246-may-19-1998/gdoc/.

Ley No. 28198 (2004) El peruano-Diario Oficial, Lima 18/marzo/2004, Ley General de Donación y Trasplante de Órganos y/o Tejidos Humanos, http://www.minsa.gob.pe/premio/archivos/legal/Ley_N28189DonacTransp.pdf.

Ley Chile (1996) http://www.leychile.cl/Navegar/?idNorma=30818&idVersion=1996-04-10&idParte.

Ley Chile (2013) http://www.leychile.cl/N?i=1051662&f=2014-06-11&p=.

Ley General de Salud, Última reforma (2012) Título Décimo Cuarto Donación, Trasplantes y Pérdida de la Vida, Diario Oficial de la Federación, México, 7 de febrero de 1984.

Lorenti A (1991) Cultivo de epidermis, *Revista Argentina de Quemaduras* **6**, 20–21.

Lorenti A (1993) Cultivo de epidermis. Primeras Experiencias en el Laboratorio de la Fundación del Quemado Dr. Fortunato Benaim, *Revista Argentina de Quemaduras* **7**, 1–2.

Lorenti A (2012) Wound healing: From epidermis culture to tissue engineering, *CellBio* **1**(2), 17–29.

Lucero JM (2012) Revista Argentina de Ortopedia y Traumatología, Vol. No. 7.

Martínez Pardo ME, Magaña DM (2007) The tissue bank at the Instituto Nacional de Investigaciones Nucleares: ISO 9001:2000 certification of its quality management system, *Cell and Tissue Banking* **8**, 221–231.

Martínez Pardo ME, Morales Pedraza J, Sánchez Ramírez O (2009) The impact of the International Atomic Energy Agency (IAEA) program on radiation and tissue banking in Mexico, *Cell and Tissue Banking* **10**, 157–165.

Martínez Pardo ME (2013) Legal system for tissue banking in Mexico, manuscript sent to the author.

Merentes E, Informe anual 2012. Proyecto regional RLA 6/062 CVIII Consolidación de Bancos de Tejidos en Latinoamérica y Radioesterilización de Aloinjertos de Tejidos. Venezuela.

Merentes Díaz E, Legal framework for tissue banks in Venezuela, manuscript sent to the author, December 2013.

Milanés CL (2012) Informe MPPS- Donación y Trasplante de Órganos y Tejidos en Venezuela. Estado actual.

Morales Pedraza J, Phillips GO (2008) The evolution and impact of the International Atomic Energy Agency (IAEA) program on radiation and tissue banking in the Latin American region, *Cell and Tissue Banking* **10**(2), 133–137.

Morales Pedraza J, Phillips GO (2009) The evolution and impact of the International Atomic Energy Agency (IAEA) program on radiation and tissue banking in Asia and the Pacific region, *Cell and Tissue Banking* **10**(2), 87–91.

Morales Pedraza J, Sánchez Noda EO, Rodríguez Cardona RL, Otero I (2009) The impact of the International Atomic Energy Agency (IAEA) program on radiation and tissue banking in Cuba, *Cell and Tissue Banking* **10**(2), 149–152.

Muscolo DL, Caletti E, Schajowicz F, Araujo ES, Makino A (1987) Tissue-typing in human massive allografts of frozen bone, *J Bone Joint Surg Am* **69**(4), 583–595.

Muscolo DL, Petracchi LJ, Ayerza MA, Calabrese ME, (1992) Massive femoral allografts followed for 22 to 36 years. Report of six cases, *J Bone Joint Surg Br* **74B**(6), 887–890.

Nather A, Morales Pedraza J (2009) The impact of the International Atomic Energy Agency (IAEA) program on radiation and tissue banking in Singapore, *Cell and Tissue Banking* **10**(2), 119–124.

Norma 241 SSA (2012) Buenas prácticas de fabricación para establecimientos dedicados a la fabricación de dispositivos médicos, Diario Oficial de la Federación, Mexico, 11 de octubre de 2012.

Ogrodnik MT (November 30, 2012) Mexico D.F., Presentation at the Forum on La tecnología de irradiación y los bancos de tejidos: retos y perspectivas.

ONT-El Modelo Español de Coordinación y Trasplantes, 2nd Edition (2008) Rafael Matezans Editor.

ONT-Venezuela, Organización Nacional de Trasplante (2006) Informe.

ONT-Venezuela, Organización Nacional de Trasplante (2013) Informe.

OPS, Washington, DC (2013) Legislación sobre donación y trasplante de órganos, tejidos y células: compilación y análisis comparado.

OPT, Organización Panameña de trasplantes, http://190.34.154.88/opt/index.aspx.

Otero I (2012) Final Report Project ARCAL CVIII — RLA 6026.

Ottolenghi CE (1966) Massive osteoarticular bone grafts: Transplant of the whole femur, *J Bone Joint Surg Br* **48**(4), 646–659.

Phillips GO, Morales Pedraza J (2003) The International Atomic Energy Agency (IAEA) program in radiation and tissue banking: Past, present and future, *Cell and Tissue Banking* **4**(2–4), 69–76.

Prezzavento G, Bolgiani A, Benaim F (1991) Aplicación del Cultivo de Keratinocitos en el Tratamiento de las Quemaduras: Presentación de Casos, *Revista Argentina de Quemaduras* **6**(3–4), 74–76.

Primer medio digital salvadoreño: elsalvador.com, February 8, 2014.

Registro Oficial No. 449 (Lunes 20 de Octubre de 2008), Constitución de la República del Ecuador, año 2008.

Registro Oficial No. 398 (4 de marzo de 2011), Ley Orgánica de Donación y Trasplante de Órganos, Tejidos y Células.

Registro Oficial No. 745 (viernes 13 de Julio de 2012) Reglamento General a la Ley Orgánica de Donación y Trasplantes de Órganos, Tejidos y Células, http://www.donaciontrasplante.gob.ec.

Registro Oficial No. 798 (Jueves 27 de Septiembre del 2012) Acuerdo Ministerial.

Rojas M (November 30, 2012) Forum: La tecnología de irradiación y los bancos de tejidos: retos y perspectivas", Mexico D.F., Presentation: Estado de los Bancos de Tejidos en Costa Rica.

Rojas M (December 2012) Final Report Project ARCAL CVIII — RLA 6026.

Sahagún Bernardino, Historia general de las cosas de la Nueva España, Editorial Alfa, México MCMLV, tomo II.

Sánchez Noda EO (2000) *Radiation and Tissue Banking*, in Phillips GO (ed.), World Scientific, Singapore, Vol. 6, pp. 315–355.

Santos J (2013) Banco de Córneas República Dominicana, Acerca de nosotros: www.thedominicaneyebank.tripod.com

Transplant Observatory on Donation and Transplantation, *Red/Consejo* Iberoamericano de Donación y Trasplantes (*RCIDT*), http://www.transplant-observatory.org/rcidt/Pages/default.aspx.

9

ESTABLISHING A LEGAL SYSTEM FOR TISSUE BANKING IN CHINA

Li Bao Xing, Zhao Yaping and Glyn O. Phillips

HISTORICAL DEVELOPMENT

Tissue banking practice in China goes back to 1949 when Professor Tu Kaiyuan first set up a bone bank in Tongji University. Allografts came mainly from using femoral heads after hip arthroplasties, osteotomy wedges after knee or hip realignment procedures and the use of bone from amputated limbs after trauma. The only processing was washing the removed tissue with water and then the bone stored in a –20–80°C freezer. There was no sterilisation procedure and the incidence of complications such as immune rejection resorption and infection was high, which limited the wide-ranging use of such allografts. This situation prevailed for almost the next 40 years.

The key initiative for further development came in 1985 when China formally joined the International Atomic Energy Agency (IAEA) programme on Radiation and Tissue Banking. A research contract (No. 4289/RB) was concluded between the China Institute for Radiation Protection (CIRP) and the IAEA programme to start a project on radiation sterilization of tissue grafts. This included a study of the biological effects of radiation on tissues, and the processing of bone, skin and amnion membrane tissue grafts for clinical applications. Subsequently, China formally joined the IAEA regional training programmes on Radiation and Tissue Banking (Morales and Phillips, 2009). With the support of IAEA, the first modern tissue bank in China, CIRP Tissue Bank, was established in

Taiyuan (Li *et al.*, 2010) and the organization became a sponsor and member of the Asia Pacific Association of Surgical Tissue Banks (APASTB). This development provided opportunities for CIRP Tissue Bank to maintain an intimate relationship with this IAEA international programme and to develop communications with other tissue banks in the region. A total of nine students (Sun Shiquan, Liu Hecheng, Li Youchen, Liu Zhanqi, Jia Xiaoling, Yin Shaofang, Zhang Yumin, Zhao Yaping and Chen Xueying) were registered in the training programmes which were held first in Sri Lanka and then in Taiyuan (1988), and they eventually graduated with a university diploma from the final training programme held at the National University Hospital of Singapore. Equipment such as a freeze-dryer and an ultra deep freezer was supplied by IAEA and with these qualified personnel and other appropriate equipment, a tissue bank in Taiyuan came into existence at a time when China was advancing the policy of reform and opening to the world. These developments have been described (Phillips, 2013). Thereafter the technique of radiation sterilisation of tissues was developed to provide tissue grafts for safe and effective clinical uses (Morales Pedraza, 2006; Nather, 2000). As a result, the CIRP Tissue Bank could move from the laboratory into the hospital, and from the nuclear system into the medical system.

It was in 1994 that the Shanxi Provincial Tissue Bank (SPTB) was established on the foundation laid by the CIRP Tissue Bank (Sun *et al.*, 2000). This non-profit organization of the Shanxi Medical Association (SMA), under the management of CIRP, developed research, processing, quality control and distribution of human tissues. It also undertook work on the public awareness of tissue donation and in expanding the clinical application of tissue grafts. SPTB also published and distributed technical information and materials to medical practitioners, and carried out related medical activities to promote the development of tissue banking and tissue grafting in China. Full details of these developments have been described (Zhang *et al.*, 2013).

DEVELOPMENT OF FORMAL REGULATORY AND LEGAL SYSTEMS

SPTB was, therefore, the first regional tissue bank in China to supply batches of tissue allograft. But at that time the government had not established any laws and regulations to administrate the development of tissue banking. The marketing of bone allograft posed a new problem in China, because there were no relevant regulations or laws for organ/tissue donation. But there was an urgent need for bone grafts, so the relevant section of the Ministry of Public Health arranged that the approval for marketing of bone products should be undertaken by the Bureau of Public Health of Shanxi Province using the same system as for the regulation of drugs. SPTB obtained a licence for processing and distribution of bone allografts; the cost for their distribution was approved by the local government in 1997. SPTB was then the only organization in China at that time to provide a series of surgical bone grafts regularly. The money charged for the grafts was used to recoup the costs for the procurement, processing, storage and distribution of tissues; the salaries for staff; and research and development.

As production grew, there was need for a parallel development of organisational and control systems. Thus, SPTB established quality systems based on IAEA guidelines and formulated the document *Technical Requirement for the Production of Tissue Allografts*, which then was registered as Enterprise Standard (Q/DS.J4001-1996) by the Bureau of Technical Surveillance of Shanxi Province. The operating guidelines were further extended in 1995 as a result of the First National Symposium on Bone Grafting held in Taiyuan and sponsored by SPTB, with the support of the relevant section of the Ministry of Public Health. In this way bone banking standards were set up and published in the *Chinese Journal of Surgery* in 1996 (Wang, 1996).

The demand for tissue graft grew rapidly so it required that the practice should be extended outside Shanxi Province. SPTB organised the First National Training Course on Bone Banking and the Second Symposium on Bone Grafting in 1995 and 2001, respectively. SPTB also provide technical support to the entities willing to establish tissue banks (Sun and Li, 2004).

After the reformation of government organizations in 1999 in China, both drugs and medical devices started to be administered by the State Drugs Administration (SDA). The Regulations for Supervision and Administration of Medical Devices were formulated and came into operation on 1 April 2000. According to this regulation, the State is responsible for the registration system which covers product manufacture of medical devices, bone allograft and other allografts which were then listed as Class III medical devices. This covers all the materials which are implanted into the human body, or used for life support or pose a potential risk to the human body. Such Class III medical devices must be inspected, approved and granted registration certificates by the drug regulatory authority operating directly under the State Council. Clinical evaluation must be conducted for Class II and Class III medical devices before they can be set into production. The drug regulatory authority under the State Council is responsible for the inspection and approval of clinical trials and the verification of all such Class III medical devices. When applying for registration of medical devices, technical standards, testing reports and other relevant information must be submitted in accordance with the provisions requested by the drug regulatory authority which operates under the State Council. The term of validity for the registration certificate of medical devices is four years. The holder of the certificate must apply for re-registration within six months of the certificate expiring.

To manufacture Class II and/or Class III medical devices requires inspection and approval by the drug regulatory authorities of the provinces, autonomous regions and municipalities directly under the central government, who will then issue a Medical Device Manufacturing Enterprise Licence. The industrial and commercial authority shall not issue a business license to enterprises which have not received a Medical Device Manufacturing Enterprise Licence. The term of validity of the Medical Device Manufacturing Enterprise Licence is five years. Upon expiration, re-inspection and license renewal shall be conducted. The State implements a mandatory safety certification system for certain Class III medical devices. A specific product list shall be established by the drug regulatory authority under the State Council, jointly with the quality and technology supervision authority. Allografts are included in this list.

Details of the allograft supply to clinics must be submitted directly to SDA for the registration and marketing clearance. The first registration step (trial register), is to investigate the safety and efficacy of the product. It belongs to product conformity. The second step (final register) is to validate the quality assurance systems of the tissue bank. This provides quality system conformity. To cover these, SDA published two Profession Standards: YY/T0287-1996: Medical Devices — Quality Management

Systems — Requirements for Regulatory Purposes and YY/T0288-1996: Quality System — Medical Devices — Particular Requirements for the Application of GB/T 19002. Both have now been replaced *by* YY/T0287-2003, based on ISO13485 and ISO13488. These standards provide the legal criteria for product conformity and quality system medical devices. SPTB became the first tissue bank to submit all materials needed for trial register according to the new regulation of SDA, and obtained approval (Li, 2009).

Now the final systems are in accordance with those set out in Biological Evaluation of Medical Devices (GB/T16886.1-1997, idt ISO10993-1:1992), including data relating to genotoxicity, cytotoxicity, carcinogenicity, chronic toxicity, local effect after implantation and irritation tests. Clinical evaluation reports from qualified hospitals with a designated number of cases are also needed.

With the preparation and application of the rapid development of tissue allografts in China, the State Food and Drug Administration considers such products should have clear regulatory level requirements. Therefore, in December 2009 it promulgated the Medical Device Good Manufacturing Practice Implementing Rules for Implantable Medical Devices (Trial), the first time it had explicitly put forward special requirements for allogeneic tissue production enterprises, so that some contents that technically were clear but not explicit in management contents now became clear.

China currently does not have an organisation similar to the American Association of Tissue Banks (AATB) or European Association of Tissue Banks (EATB). The Ministry of Health released a WS 324-2010 Standard for Human Musculoskeletal Tissue Bank in 2010, in which the comprehensive quality management requirements of bone tissue banking was presented; its purpose is to regulate behaviours of organisations that originated in the hospitals, and that prepared allograft products to be only used within the hospital. The State Food and Drug Administration (SFDA) also at the end of 2009 promulgated three relevant industrial standards: YY/T 0513.1-2009 Allogeneic bone grafts Part I: The basic requirements of bone tissue bank; YY/T 0513.2-2009 Allogeneic bone grafts Part II: Deep frozen allogeneic bone grafts and freeze-dry allogeneic bone grafts; and YY/T 0513.3-2009 Allogeneic bone grafts Part III: Demineralized allogeneic bone grafts. The principle of Part I is similar to that of the Ministry of Health standards, and provides basic quality system requirements for a bone tissue bank, while Parts II and III proposed performance requirements for various allografts bone products. Two of the three relevant industrial standards were written by SPTB.

Following the changed regulations of government, SPTB changed from being a non-profit organisation to becoming a stock company (Shanxi OsteoRad Biomaterial Co., Ltd.) in November 1999 as required, in order to be licensed from SDA for marketing of bone allograft to the hospitals. The income has been sufficient to achieve self-dependence and provide a stable financial situation. The trial register was submitted to SDA, and it obtained the first approval of its productive registration from SFDA in 2001. In April 3, 2014, CFDA (SFDA was renamed in 2013 as FDA) released a regulation for the supervision and administration of medical device manufacturers. This regulation classified medical device manufacturers into four classes according to the level of risk of their productions. The manufacturers of allografts and xenografts were designated as Fourth Class, the highest class. The government regulation of allografts manufacture will be more stringent.

The present Shanxi OsteoRad Biomaterial Co., Ltd./SPTB has 35 employees of which there are four professors (two are post-graduate mentors), four associate researchers, four senior laboratory assistants, three high-grade technicians and 14 postgraduates (PhD and Master's degree). The OsteoRad Corporation owns 12 national patented products and techniques, was awarded a first grade prize from the Ministry of Science and Technology Advancement programme and one second grade prize, and has completed two military research contracts and five from provincial or national sections. Its facilities provide 1,800 square metres of production area, and houses 500 square meters of a ten thousand-class clean room. The facilities include advanced processing equipment and facilities with a built-in series of laboratories for pathology, chemistry, microbiology, biomechanics, electron microscopy, animal dissection and tissue engineering.

OTHER TISSUE BANKS DEVELOPING LEGAL SYSTEMS TO GAIN NATIONAL APPROVAL THROUGHOUT CHINA

The time taken by the legal system to move from the original association of SPTB with the IAEA to when it gained approval to produce and distribute tissue grafts throughout China took about 20 years. This was mainly due to the lack of technical systems to identify tissue grafts and to legally control the procurement, processing, packing, distribution and clinical use. Now this pioneer bank has gained national acceptance throughout China and has served as a model for further approval of other tissue banks.

Even so, there are only a very limited number of tissue banks which have approval to produce and distribute tissue outside their hospital and throughout China. Other well-established tissue banks include the following:

- Wuhan, Lianjie Company/Osteolink Biomaterials Co. Ltd., Hubei http://www.osteolink.com.cn/index.asp;
- Beijing, Datsing/Biotech Company, http://www.datsing.net/dts_bio_index.asp.

Other tissue banks that have more recently been approved are the following:

- Bejing, Xinkangchen Company, based on the bone tissue bank in First Affiliated Hospital of PLA General Hospital;
- Shanghai, Anjiu Company, based on the tissue bank in Ninth Affiliated Hospital of Shanghai Jiao Tong University bone tissue bank, http://www.ajbio.com;
- Tianjin, Jinxingda Company (this is essentially a cancer facility);
- Beijng, Yunkanghengye Company (2012-11 approved).

REFERENCES

Li BX (2009) Shanxi Provincial Tissue Bank: 20 years of experience, *Chin J Repar Reconstr Surg* **23**(5), 547–548.

Li D, Meng GL, Liu M, Jin J, Liu Y, Wang Z, Liu J, Hu YY(2010) Multi-variety bone bank in China, *Cell Tissue Bank* **11**, 233–240.

Morales Pedraza J (2006) The International Atomic Energy Agency (IAEA) program in radiation and tissue banking: A successful program for developing countries, *Cell Tissue Bank* **7**(4), 237–258.

Morales Pedraza J, Phillips GO (2009) The evolution and impact of the International Atomic Energy Agency (IAEA) program on radiation and tissue banking in Asia and the Pacific region, *Cell Tissue Bank* **10**, 87–91.

Myint P, Wondergem J, Pynda Y, Phillips GO (2013) The use of the ionization technique for tissue sterilisation: The International Atomic Energy Agency (IAEA) Experience, *Cell Tissue Bank* **14**, 159–165.

Nather A (2000) Regional developments in tissue banking in the Asia and the Pacific region: Ethical, religious and cultural factors legal and other regulatory aspects, *Radiation and Tissue Banking*, World Scientific, Singapore, pp. 219–234.

Sun SQ, Li BX (2004) Bone bank technology and operating standard, *Chin J Orthop* **24**(10), 638–640.

Sun SQ, Li BX, Li YC (2000) Establishment and development of Shanxi Provincial Tissue Bank, *Radiat Prot Bull* **20**(1), 41–44.

Wang GF (1996) Establishment and perfect the regulation system of tissue bank is imperative, *Chin J Surg* **34**(8), 451–452.

Zhang YM, Li BX (2006) Development of modern tissue banks and quality control, *Chinese J Reparative and Reconstructive Surgery* **20**(11), 1134–1137.

Zhang YM, Wang JR, Zhang NL, Liu XM, Zhou M, Ma SY, Yang T, Li BX (2013) Rapid development of tissue bank achieved by the International Atomic Energy Agency (IAEA) Tissue Banking Programme in China, *Cell Tissue Banking* DOI 10.1007/s10561-013-9393-x.

10

INDEPENDENCE FOR WALES FROM THE REMAINDER OF THE UNITED KINGDOM: A CHANGE FROM INFORMED CONSENT TO PRESUMED CONSENT

Glyn O. Phillips

INTRODUCTION

Wales lost its legal independence in 1536 when the Act of Union with England was enacted in London. We have continued to resent the situation and when opportunities offer themselves our national desires emerge. Welsh Nationalist members have been elected from Wales to the British Parliament and continue to remind our English neighbours that we are different. Wales was the language of court princes, almost a thousand years before the time of Chaucer — the father of English literature. Welsh has been and continues to be a living language ever since the Romans left these shores about the middle of the fourth century. Then the old Brythonic language became truly Welsh and can be identified with the Cymraeg spoken today. This independence manifests itself in many ways, and although many have lost the language, they continue to be fiercely Welsh when called upon — as anyone who has been present when Wales plays England at rugby in Cardiff or Twickenham will testify!

The strongest recent expression of this growing independence came in 1997 when, in a referendum, Wales voted for devolution from England which was enacted through the Government of Wales Act 1998. There is now a functioning Assembly/Senedd where laws specific to Wales can be identified and implemented as, for example, in the conduct of finance, education, the Welsh language and agriculture.

Now perhaps the most unique result of this independence is the passing, with the blessing of the Queen and the approval of the United Kingdom Parliament, of the Human Transplantation (Wales) Bill. This Act became law in Wales in 2013. Human tissue donation throughout the United Kingdom has hitherto followed the informed consent principle and will continue to do so in those parts of the United Kingdom outside Wales. But the historic step has been taken now to change this in Wales where the principle of presumed consent will be the law. It is a historic change which once again emphsises the strength of indpendencce in Wales. Thus as the Health Minster guiding the legislation declared, "Tuesday July 2nd was a landmark day for Wales, and I expect the rest of the UK will watch with great interest when the legislation is implemented in 2015." He added in a published statement that "the people of Wales should be proud that we are the first nation in the UK to take this step".

THE SCOPE OF THE HUMAN TRANSPLANTATION (WALES) BILL

The Act sets out the legal framework for consent to the donation of organs and tissues in Wales for transplantation. It introduces a soft opt-out system for procuring organs from the deceased and tissue donation:

> "A soft opt-out system is one where consent to the removal and use of organs and tissues for transplantation is deemed as having been given unless the deceased objected during their lifetime, and where the next-of-kin will be involved in the decision-making process."

The objective is to increase the number of organs and tissues available for transplantation and to reduce the number of people dying whilst waiting for a suitable organ, and to improve the lives of others. The expectation is that donation rates might increase by 25–30% under an opt-out system.

The draft Bill was published first on 18 June 2012 and was the basis of a wide consultation which ran from 7 November 2011 until 31 January 2012. 1,234 responses were received and as a result the present draft Bill was published. A consultation report was produced on 8 March 2012. Full details can be obtained from the Wales government website: (www.cymru.gov.uk). The Human Transplantation (Wales) Act received Royal Assent on 10 September 2013. This law will come fully into effect on 1 December 2015. Until then, the current opt-in system will remain in place. Although the concept of presumed consent is the underlying principle for the Act, there has been a tactical compromise introduced to attempt to divert possible objections by adopting the principle of a "soft opt out".

Only some concepts of the new Law will be noted here, and for full details the Bill published by the Wales Assembly government should be consulted as on their website. The main provisions within the Act are to

(a) impose a duty on the Welsh Ministers to promote transplantation;
(b) provide that certain activities carried out in Wales for the purpose of transplantation ("transplantation activities") are lawful if done with consent;
(c) set out how consent is given to transplantation activities, including the circumstances in which consent is deemed to have been given in the absence of express consent;
(d) make it an offence for transplantation activities to be carried out in Wales without consent;
(e) make consequential amendments to the Human Tissue Act 2004 including in relation to a code of practice issued under that Act which —
    (i) gives practical guidance to persons carrying out transplantation activities, and to
    (ii) lay down the standards expected in relation to the carrying out of such activities, including how consent is to be obtained.

However, the intention can be illustrated in the following example noted in the Act.

Illustration of the Meaning of "Express Consent"

The person is deemed to have given his or her consent in relation to the activity unless

1. The person is alive — when the person's consent is given
2. The person has died and a decision of the person to consent, or not to consent, to the activity was in force immediately before his or her death — when the person's consent is given.

3. The person has died, case 2 does not apply and the person had appointed a person or persons to deal with the issue of consent in relation to the activity — when the consent is given by the person or persons appointed to deal with the issue of consent in relation to the activity.

Special provisions are provided for

(a) a body which has been imported into Wales,
or
(b) relevant material which —
   (i) has been imported into Wales, or
   (ii) has come from a body which has been so imported.

Special provisions are set out in the case of person who is a child or has died a child and for adults who lack the capacity to consent.

There are provisions to enable a person to appoint a representative to express his or her wishes as follows:

- An adult may appoint one or more persons to represent the adult after his or her death in relation to express consent.
- An appointment may be general or limited to consent in relation to such one or more transplantation activities as may be specified in the appointment.
- An appointment may be made orally or in writing.
- An oral appointment is only valid if made in the presence of at least two witnesses present at the same time.

Of particular relevance to tissue banking activities where dead tissue is still of value for transplantation is the following clause:

> Where part of a body of a deceased person lying in hospital, nursing home or other institution is or may be suitable for use for transplantation, it is lawful for the person having the control or management of the institution:

(a) to take steps for the purpose of preserving the part for use for transplantation,
and
(b) to retain the body for that purpose.

As a result of the introduction of this Act in Wales, there are consequential amendments to the Human Tissue Act 2004 which should be noted.

MOTIVATION AND RESPONSE

Why was this legislation given priority in Wales? As noted by the Minister, about three people in Wales die every month waiting for a suitable organ. The Bill is designed to address this tragedy. Currently about a third of donations to the deceased come from people who said they wanted to be donors and joined via the Organ Donor Register. This means that two-thirds are, therefore, from people who were not on the Organ Donor Register at all. The hope is that the implementation of the Act will initiate a new type of consensus and open up discussion within families on thoughts and feelings about donating organs and tissues after death.

The Welsh Assembly government deserve nothing but the highest praise for taking and diligently pursuing this controversial topic. It leads the way within the sovereign countries which make up the United Kingdom and throughout Europe where only Belgium holds to the same prinicple.

Yet at the moment of this triumph there is an anomaly. There is no tissue bank in Wales. The live organs naturally will be sent and used wherever they are needed in the United Kingdom or elsewhere where there is a suitable recipient — which on the basis of population will almost certainly be outside Wales. This act of altruism on the part of the people of Wales is praiseworthy and typical of their international perspective.

But it is ironic, on the other hand, that there is now no human tissue bank in Wales since the functioning tissue bank in Wrexham/Oswestry was purchased by the national blood transfusion service and transferred to Liverpool. The logistics are now so ineffective that there is not even a tissue collection service from deceased donors in South Wales because of a perceived problem in travelling 200 miles from Liverpool to South Wales.

In response to my enquiry about this unsatisfactory situation, the Minister for Health and Social Services in Wales wrote:

> I am aware of this situation since we gave specific attention to this matter towards the end of 2009. Then the UK Blood Transfusion Service

> prepared a report on a service for tissues in Wales which would be controlled centrally (this presumably means centrally in the UK and outside Wales).
>
> The UK Blood and Transfusion Service has three centres to process tissues — Liverpool, Leeds and London. Liverpool collects the available tissues from North Wales. In South Wales only specific tissues are collected. After considering all relevant factors it was decided that it would be too costly and unnecessary to provide a service for Wales. The hospitals in Wales can now still purchase tissues at a reasonable cost from outside providers.

The Minister added in a note addressed to me: "In practice there is no general shortage of tissues in the United Kingdom at the present time and the concerns expressed by Professor Phillips are unlikely to arise."

This contradicts the reasoning used to justify the unique tissue donation system now passed in law in Wales only, which was based to a large extent on there being a shortage of tissues available for human transplantation. It is ironic that while there is a tissue bank for animals functioning effectively in Wales, there is not one for people.

11

GERMANY: REGULATION OF TISSUES AND TISSUE BANKING — A SPECIAL CASE

Peter Hergert and Ralf R. Tönjes

INTRODUCTION

Situation Before the EU Tissue Directive

In Europe, from the beginning of the new millennium, tissue products for clinical use have received more and more attention from regulatory authorities. They noticed that there was an increasing need and demand for such products leading to increasing exchange of these products within the European Union (EU). At that time, tissue products were largely unregulated and harmonised at a European level. The first EU tissue directive was then adopted in 2004 (Directive 2004/23/EC).

In Germany, too, activities and products of tissue banks were unregulated for a long time. It was essentially left up to medical doctors and their scientific societies to provide high-quality and safe tissue products and to regulate the use of such products.

The competent German health authorities noticed this state of affairs in the mid-1990s. As a result, it was considered whether the most common tissue products were subject to the provisions governed by the established drug law in Germany. But at that time, the drug regulatory authorities were not familiar with tissue products and the stakeholders in that field. For that reason, the authorities first created an overview for themselves about the type and the scope of the existing tissue establishments and their products and activities. Therefore, they called for a public survey published in the *Bundesanzeiger* (Federal Gazette) of the federal German governmental administration.

As a result, it was concluded that in Germany the tissue products used at that time fell within the definition of medicinal products and would therefore be subject to regulation via the drug law. In particular, the activities of tissue banks should be covered by manufacturing authorisations and the tissue products themselves by marketing authorisations. An immediate implementation of these requirements, however, would have led to a tremendous burden for tissue banks and possibly unacceptable shortages in transplantation medicine, surgery and orthopaedics. For this reason, the German authorities granted long transitional periods to give tissue banks the chance to adapt their activities and products to the specifications laid down in the drug law step by step.

Concurrently the EU tissue directive caught up with this (inner German) evolution. The inclusion of tissue products into the system of drug regulation had already begun in Germany, but it needed to be continued within the meaning of the new European provisions.

This explains why the structure of the present German tissue law is more complex than that of the EU and of most other EU member states. The German tissue law extends across several fields of law.

Transposition of EU Tissue Directive into German Law (an Overview)

Since the regulation of tissue products under the drug law in Germany had already started, the steps already taken were not entirely withdrawn in Germany. Thus a uniform legal framework could not be created for these products.

Directive 2004/23/EC (EU tissue directive) was transposed into German law in two steps. First, the corresponding section of the drug law governing Good Manufacturing Practice was modified such that tissue establishments (particularly tissue banks and test laboratories) could be regulated under this law.

The second (more extensive) step was to carry out modifications in several other fields of law. First, the transplantation law, which up to that point was only applicable to organs, was extended to the transplantation of tissues and tissue products. Second, special provisions for tissue products were included into drug law so that the system of drug regulation could in principle be applied to these products as required by the EU tissue directive. Finally, the transfusion law was amended, since stem cell preparations from human blood were to be regulated within this legal framework. But on a European level they fall within the scope of the EU tissue directive.

LEGAL FRAMEWORK

European Law

Only a brief overview of the European legal framework for tissue products and tissue establishments will be given since it is dealt with in another chapter. The focus of this chapter will be on a description of the German law.

Directive 2004/23/EC lays down the fundamental requirements which member states must fulfil for regulating tissue products and monitoring tissue establishments. The European law makers used as a model Directive 2002/98/EC for the regulation of blood products and blood banks which had been introduced a few years before.

In delegated directives the European law maker set out detailed technical requirements for the donation, procurement and testing of tissues (Directive 2006/17/EC), traceability and notification of serious adverse reactions and events (Commission Directive 2006/86/EC) and conditions for inspections and qualifications of inspection officials (Commission Decision 2010/453/EU).

German Law

Drug law

Under German law, human tissues and preparations from human tissues are considered medicinal products according to Section 4(30) of the German Medicinal Products Act [MPA] (Arzneimittelgesetz [AMG]). Thus, in principle they fall within the scope of German drug law. However, the drug law lays down a number of special stipulations for tissue preparations. Parts of these deviate greatly from those for other medicinal products and they are adapted to the special characteristics of tissue preparations and tissue establishments.

Permission for the procurement of tissues and laboratory testing (Section 20b MPA)

The first step towards using a tissue preparation within a medical treatment is the procurement of the raw tissue material. Facilities for collecting such tissues in Germany (procurement establishments) require an authorisation pursuant to Section 20b MPA to do this. Establishments performing the laboratory testing required for such tissue procurement also require an authorisation pursuant to Section 20b MPA.

An authorisation pursuant to Section 20b MPA for the procurement and/or testing of tissues is granted by the authority in whose territory the establishment is located. The responsible authorities are part of the administration of the German federal states (Länder). A list of these competent local authorities can be found on the website of the German Zentralstelle der Länder für Gesundheitsschutz bei Arzneimitteln und Medizinprodukten [ZLG] (Central Authority of the Länder for Health Protection with regard to Medicinal Products and Medical Devices) at its website, https://www.zlg.de/arzneimittel/deutschland/laenderbehoerden.html.

The respective establishment in principle has a legal right to being granted an authorisation pursuant to Section 20b MPA unless

1. There is no appropriately trained (usually medical) member of staff with the required professional experience;
2. The other involved members of staff are not sufficiently qualified;
3. The premises are not suitable for the respective tissue procurement or lab testing; or
4. The performance of the tissue procurement or lab testing is not guaranteed pursuant to the state of art of medical science.

An authorisation pursuant to Section 20b MPA is granted only for one particular site. For multiple sites, a procurement facility requires separate authorisations. Furthermore, the authorisation pursuant to Section 20b

MPA is not granted for procurement of human tissue and/or lab testing in general, but the permitted types of tissues and tests are explicitly mentioned in the authorisation.

A separate authorisation pursuant to Section 20b MPA can be omitted if the procurement and/or lab testing is performed for another establishment based on a contractual agreement, and if the latter establishment has been granted a manufacturing authorisation pursuant to Section 13 MPA or a processing authorisation pursuant to Section 20c MPA (as below). In this instance, the authorisation for the procurement and/or lab testing of the tissues can be included in the manufacturing authorisation or the processing authorisation. The communication with the competent authority will then usually be taken over by the (probably more experienced) holder of the manufacturing authorisation or processing authorisation, which is in practice a considerable relief from workload for the procurement establishment or testing laboratory.

To receive an authorisation pursuant to Section 20b MPA, the applicant is obliged to observe the regulations of Section 5a of the "Arzneimittel- und Wirkstoffherstellungsverordnung" [AMWHV] (Medicinal Products and Active Substance Manufacturing Ordinance). These contain special stipulations for establishments which procure, process and/or test tissues.

Authorisation for the processing, preservation, testing, storage and placing on the market of tissues and tissue preparations (Section 20c MPA)

Establishments which process, preserve, test, store or place on the market tissues and tissue preparations require an authorisation pursuant to Section 20c MPA. The permission pursuant to Section 20c MPA is equivalent to the manufacturing authorisation for other medicinal products (Section 13 MPA, Art. 40 Directive 2001/83/EC).

The requirements for granting an authorisation pursuant to Section 20c MPA are less strict than those for granting a manufacturing authorisation pursuant to Section 13 MPA. The establishment has to fulfil the following requirements:

1. Employ a sufficiently experienced expert responsible for the treatment or processing, preservation, testing, storage or placing on the

market of the tissue or tissue preparation in conformity with the applicable legal provisions;
2. Ensure that the other involved members of staff are sufficiently qualified;
3. Provide suitable premises and facilities for the activities intended;
4. Guarantee that treatment or processing, including labelling, preservation, storage, and testing are performed in compliance with the state of art of science and technology; and
5. Establish a quality system in compliance with the principles of Good Practice, and keep this system up to date.

The core of these requirements is the adherence to Good Practice which tissue establishments have to consider. From this point of view, Good Practice for tissue preparations represents the equivalent of Good Manufacturing Practice for other medicinal products. More precise information concerning this is provided in Section 5a of the AMWHV.

If the tissue establishment fulfils the abovementioned requirements, there is in principle a legal right to be granted an authorisation pursuant to Section 20c MPA.

An authorisation pursuant to Section 20c MPA, however, can only be granted for tissue preparations which are not treated or processed by means of an industrial process. Furthermore, the essential procedures of treatment and processing have to be well known in the European Union (so-called "classical tissue preparations"). For tissue preparations that do not fulfil these requirements, a manufacturing authorisation pursuant to Section 13 MPA is required. This also means that the principles of Good Manufacturing Practice for medicinal products have to be met. For such cases Section 3 AMWHV lays down special provisions for products of human origin — including tissue preparations.

An authorisation pursuant to Section 20c MPA is granted only for a particular site and for particular tissues and tissue preparations.

The German local competent authorities for the granting of an authorisation pursuant to Section 20c MPA are the authorities of the Land (German Federal State) where the respective establishment is seated (see also https://www.zlg.de/arzneimittel/deutschland/laenderbehoerden.html).

The authorisation is generally granted only following an official inspection of the site.

Before deciding on the authorisation, the local competent authority consults the Paul Ehrlich Institute, which is the competent authority of the German federal administration. In practice, the consultation is performed by involving representatives of the Paul Ehrlich Institute in the abovementioned inspection.

Authorisation for the placing on the market of tissue preparations (Section 21a MPA)

The authorisations pursuant to Sections 20b and 20c MPA refer to the activities of a tissue establishment, i.e., procurement, laboratory testing, processing, etc. The finished tissue preparation also requires a marketing authorisation pursuant to Section 21a MPA. This authorisation is the equivalent of the marketing authorisation for other medicinal products (Section 21 MPA, Art. 6 of Directive 2001/83/EC). However, the conditions for this authorisation are adapted to the special characteristics of tissue preparations, and the authorisation procedure is simplified considerably as compared to a marketing authorisation procedure for other medicinal products.

An authorisation pursuant to Section 21a MPA is required only if the tissue establishment places its products on the market, i.e., in particular, if it intends to pass them to other institutions. A tissue bank which is integrated into a hospital and provides only its own hospital with its products is not considered as placing its products on the market, and for this reason, does not require an authorisation pursuant to Section 21a MPA. In practice, competent authorities assume that this is the case if tissue preparations are handed out only within the same site as laid down in the authorisation pursuant to Section 20c MPA.

As in the case of Section 20c MPA, an authorisation pursuant to Section 21a MPA can only be granted for the so-called “classical tissue preparations”. Tissue preparations treated or processed by means of an industrial process, or whose essential treatment or processing procedure is not (yet) well known in the EU require an authorisation pursuant to Section 21 MPA. In particular, this will then also require an application in

Common Technical Document (CTD) format and the submission of non-clinical and clinical data.

The documents to be submitted with an application for an authorisation pursuant to Section 21a MPA are described in Section 21a (2) and (3) MPA. The listing contained in that section, however, is phrased in very general terms. It is therefore recommended that applicants use the forms provided by the Paul Ehrlich Institute (see http://www.pei.de/DE/infos/pu/genehmigungen/gewebe-21-a-amg/01-antrag-genehmigung/antrag-genehmigung-node.html). That way, it can be assured that all important information and documents are provided with the application. Queries and/or objections from the Paul Ehrlich Institute will then usually concentrate on scientific aspects which will reduce the time required for the authorisation procedure to a minimum.

The Paul Ehrlich Institute will grant the authorisation if the application documents show that the tissue preparation meets the state of art of scientific knowledge, fulfils its envisaged function, and proves to have a positive risk-benefit ratio (Section 21a (6) MPA).

Duties after receiving the authorisations pursuant to Sections 20b, 20c and 21a MPA

Modifications of the information and/or data on which the authorisations pursuant to Sections 20b, 20c AMG have been granted (e.g., modifications of the rooms or the equipment used) or the authorisations pursuant to Section 21a AMG have been granted (e.g., modifications of the processing procedure or product testing) must be reported to the appropriate competent authority before implementation. Some modifications may only be implemented after the competent authority has given its consent.

Pursuant to Section 63i MPA, a tissue bank is obliged to document extensively all suspected serious adverse events or reactions in connection with their tissue preparations as well as any recalls, and report them to the competent authorities.

Imports

As a matter of course, tissue establishments outside Germany which are located in a member state of the EU or the European Economic Area (EEA) at first require all the necessary permits or authorisations in accordance with the laws of "their" member state.

Furthermore, if a third-country tissue establishment wishes to place its tissue preparations on the German market, it will require a certificate pursuant to Section 21a (9) MPA by the Paul Ehrlich Institute. The procedure for obtaining such a certificate is modelled on the above described authorisation procedure pursuant to Section 21a MPA. Permits or authorisations of the "exporting member state", however, are taken into account so that the documentation obligations can be considerably reduced compared to a usual application pursuant to Section 21a (1) MPA.

On the other hand, for tissue preparations imported to Germany from countries outside the EU or the EEA, the importing tissue establishment requires an authorisation pursuant to Section 72b MPA. The conditions for receiving such an import authorisation are essentially equivalent to the above described conditions for receiving authorisations pursuant to Sections 20b, 20c MPA. Furthermore, an authorisation pursuant to Section 21a MPA is required to place the tissue preparations on the market, i.e., to sell them to the end-consumer facilities.

Transplantation law

In Germany, the technical and medical/biological requirements for tissue preparations laid down in Directive 2004/23/EC and the appropriate implementing or delegated Directives (see the above Section 2.1) are regulated by the German Transplantation Act (Transplantationsgesetz [TPG]) and the German Tissue Ordinance based on the TPG (TPG-Gewebeverordnung [TGP-GewV]).

These in particular regulate the requirements for the procurement of tissue from dead and live donors (Sections 2 and 3 TPG). Tissue banks are obliged to comply with the current state of art of medical knowledge (Section 8d(1) TPG). Besides, special documentation and reporting obligations are laid down (Section 8d(2) and (3) TPG).

Tissue establishments seated in Germany must have themselves registered in a public database pursuant to Section 8f TPG, which is managed by the German Institute for Medical Documentation and Information (Deutsches Institut für Medizinische Dokumentation und

Information [DIMDI]). The data to be submitted and registered are laid down in the German Ordinance on Registration of Tissue Establishments based on the TPG (TPG-Gewebeeinrichtungen-Registerverordnung [TPG-GewRegV TPG]).

Many legal provisions require the adherence to the current state of art of scientific/medical knowledge in connection with the procurement and transfer of tissues. This scientific/medical knowledge, however, has not been written down anywhere accurately and is not legally binding. This is due to the nature of the subject matter, since scientific knowledge is subject to constant change. For this reason, the law makers have left it up to the self-management of the German Physicians' Organisation (Deutsche Bundesärztekammer [BÄK]) to lay down the generally recognised state of art of medical science in guidelines (Section 16b TPG).

Even though these guidelines are not legally binding, a tissue establishment which observes them can assume that its procedures and processes will not be judged unfavourably by the authorities. A facility that deviates from these guidelines may have to justify why the chosen procedures and processes still adhere to the state of art of medical knowledge.

At the time of the deadline for this publication, one guideline has been adopted pursuant to Section 16b TPG (Guideline on the Procurement of Donor Cornea Material and the Management of a Cornea Bank — Richtlinie zur Gewinnung von Spenderhornhäuten und zum Führen einer Augenhornhautbank).

Clinical trials

Since tissue preparations are medicinal products according to German law, the legal provisions for clinical trials of medicinal products also apply in principle to tissue preparations.

The medical use and the safety of classical tissue preparations are largely proven and extensively published, which means that as a rule, clinical trials are not performed with classical tissue preparations. In spite of this, cases are possible in which, for instance, the use of a classical tissue preparation shall be clinically tested against another medical measure.

In such instances it may be required to apply for a positive opinion from the competent ethics committee(s) as well as for an authorisation for a clinical trial by the Paul Ehrlich Institute (Sections 40ff MPA in combination with the German Ordinance on Good Clinical Practice — GCP Ordinance (GCP-Verordnung [GCP-V]).

In future, the provisions of the European Regulation (EU) No 536/2014 will apply to clinical trials.

Special cases: Tissue products regulated differently

Human tissues and cells are sometimes also used for products or treatments for which the abovementioned principles and legal frameworks are not applicable.

Advanced therapy medicinal products

In particular, advanced therapy medicinal products (ATMP) such as somatic cell therapeutics, tissue-engineered products and gene therapeutics have to be mentioned in this context. Somatic cell therapeutics and tissue-engineered products by definition contain human tissue and/or cells. Thus, they also fall under the definition of tissue preparations due to Section 4(30) MPA (see above 2.2.1). Gene therapeutics may contain human cells.

These medicinal products are not regulated under the special provisions for tissue preparations as described above but in accordance with the general provisions of the Medicinal Products Act and the other drug law. In addition, special provisions for ATMPs themselves do apply — in particular the European Regulation (EC) No 1394/2007, but also several provisions in the German drug law.

The procurement and laboratory testing of the starting tissues or starting cells for ATMPs, however, are still subject to the above described provisions of the transplantation law.

Stem cell products

Stem cell preparations are derived from either peripheral blood, cord blood or bone marrow from a donor. There are no significant differences between the preparation and medical use of stem cell preparations from peripheral blood or cord blood and those from bone marrow.

In accordance with the definitions of the German Medicinal Products Act, stem cell preparations from peripheral and cord blood are blood preparations (Section 4(2) MPA), and stem cell preparations from bone marrow are tissue preparations (Section 4(30) MPA). This means that different special pharmaceutical provisions apply for stem cell preparations, depending on their origin.

Moreover, stem cell preparations from bone marrow are regulated under the TPG (and other provisions of the transplantation law). Stem cell preparations from peripheral and cord blood, however, fall under the German transfusion law, in particular the Transfusion Act (Transfusionsgesetz [TFG]). At European level, on the other hand, all stem cell preparations fall in the scope of the Tissue Directive (Directive 2004/23/EC), since they are explicitly exempted from the scope of the Blood Directive (Art. 2(4) of Directive 2002/98/EC).

Considering the fact that one establishment that works with stem cell preparations usually processes all types of stem cell preparations (in Germany mostly blood banks), the separation into two German fields of law has led to various problems in the past which are not yet entirely resolved.

Germ Cells

According to German law, human germ or reproductive cells (sperm and ovarian cells) by definition are neither tissue preparations nor medicinal products (Section 4(30) second sentence MPA). For this reason, the provisions of the German drug law are not applicable for germ cells.

However, germ cells are tissues in the meaning of the German transplantation law. The activities of an establishment which procures, processes and stores germ cells therefore fall under the provisions of the TPG and the TPG-GewV. The latter contains special provisions for germ cells intended for the use within measures of assisted reproductive technologies (ART) (Section 6 TPG-GewV and Annex 4 of TPG-GewV).

Medical Devices Containing Tissues or Cells

Medical devices are sometimes combined with human tissues or cells. The medical suitability of these products often is only derived from this special combination.

The legal classification of such products as medical devices, tissue preparations, ATMPs, or other medicinal products can cause major difficulties and is often the subject of time-consuming negotiations between the involved companies, tissue establishments and competent authorities. Due to the complexity of these questions and the variety of possible products, it is not possible in this chapter to give guidance for the classification of such products which would be suitable for practice.

However, the first question is always if the product shall be (primarily) regulated under the laws governing medicinal products or medical devices. In practice such classifications are made with the aid of the guideline of the EU Commission document MEDDEV 2. 1/3 rev 3. The MEDDEV guidelines are not legally binding, but have a high status in legal practice and are generally accepted.

FINAL REMARKS

In Germany, no uniform legal framework for tissues and tissue preparations based on the structures of the European tissue directives has been created. Tissues, tissue preparations and tissue establishments are governed by provisions which extend over several legal fields, in particular the drug law and the transplantation law.

For this reason, competent technical and regulatory advice is essential for inexperienced institutions that intend to place tissue preparations on the German market or intend to establish facilities in Germany.

However, competent German authorities have found solutions for a large number of initial questions and issues to regulate and monitor tissues and tissue preparations as well as tissue establishments in both a professional manner and in complying with the applicable laws.

At the level of the Länder (German Federal States), supervisory experts hold regular meetings in a working party of the ZLG to discuss questions from the areas of tissue and tissue establishments and agree upon uniform procedures.

The Paul Ehrlich Institute offers scientific advice in its field of competence before submitting an application. Regulatory issues, too, in connection with authorisations and post-marketing control of tissue preparations, can be addressed here.

REFERENCES

Arzneimittelgesetz — AMG (Medicinal Products Act — MPA) in the version of the announcement of 12 December 2005 (Federal Law Gazette I, p. 3394), last amended by Article 2a of the Act of 27. March 2014 (Federal Law Gazette. I, p. 261).

Commission Directive 2006/86/EC of 24 October 2006 implementing Directive 2004/23/EC of the European Parliament and of the Council as regards traceability requirements, notification of serious adverse reactions and events and certain technical requirements for the coding, processing, preservation, storage and distribution of human tissues and cells (OJ L 294 of 25 October 2006, p. 32).

Commission Decision 2010/453/EU of 3 August 2010 establishing guidelines concerning the conditions of inspections and control measures, and on the training and qualification of officials, in the field of human tissues and cells provided for in Directive 2004/23/EC of the European Parliament and of the Council (notified under document C(2010) 5278) (OJ L 213 of 13 August 2010, p. 48).

Directive 2001/83/EC of the European Parliament and of the Council of 6 November 2001 on the community code relating to medicinal products for human use (OJ L 311 of 28 November 2001, p. 67), last amended by Directive 2012/26/EU of the European Parliament and of the Council of 25 October 2012 (OJ L 299 of 27 October 2012, p. 1).

Directive 2002/98/EC of the European Parliament and of the Council of 27 January 2003 on setting standards of quality and safety for the collection, testing, processing, storage and distribution of human blood and blood components and amending Directive 2001/83/EC (OJ L 33 of 8.2.2003, p. 30), amended by Regulation (EC) No 596/2009 of the European Parliament and of the Council of 18 June 2009 (OJ L 188 of 18 July 2009, p. 14).

Directive 2004/23/EC of the European Parliament and the Council of 31 March 2004 on setting standards of quality and safety for the donation, procurements, testing, processing, preservation, storage and distribution of human tissue and cells (OJ L 102 of 7.4.2004, p. 48), amended by Regulation (EC) No 596/2009 of the European Parliament and the Council of 18 June 2009 (OJ L 188 of 18 July 2009, p. 14).

Directive 2006/17/EC of the Commission of 8 February 2006 implementing Directive 2004/23/EC of the European Parliament and of the Council as regards certain technical requirements for the donation, procurement and testing of human tissues and cells (OJ L 38 of 9 February 2006, p. 40), amended by Commission Directive 2012/39/EU of 26 November 2012 (OH L 327 of 27 November 2012, p. 24).

Gesetz über die Spende, Entnahme und Übertragung von Organen und Geweben (Transplantationsgesetz — TPG) (Act governing the donation, procurement and transfer of organs and tissues [Transplantation Act]) in the version of the announcement of 4 September 2007 (Federal Law Gazette I, p. 2206), last amended by Article 5d of the Act of 15 July 2013 (Federal Law Gazette I, p. 2423).

Medical Devices: Guidance document — borderline products, drug-delivery products and medical devices incorporating, as an integral part, an ancillary medicinal substance or an ancillary human blood derivative (MEDDEV 2. 1/3 rev 3) (http://ec.europa.eu/health/medical-devices/files/meddev/2_1_3_rev_3-12_2009_en.pdf).

Regulation (EC) No 1394/2007 of the European Parliament and of the Council of 13 November 2007 on advanced therapy medicinal products and amending Directive 2001/83/EC and Regulation (EC) No 726/2004 (OJ L 324 of 10 December 2007, p. 121), amended by Regulation (EU) No 1235/2010 of the European Parliament and of the Council of 15 December 2010 (OJ L 348 of 31 December 2010, p. 1).

Richtlinie zur Gewinnung von Spenderhornhäuten und zum Führen einer Augenhornhautbank (Guideline on the Procurement of Donor Cornea Material and the Management of a Cornea Bank) of 14 February 2014 (Deutsches Ärzteblatt 111 (31–32) 4 August 2014, pp. 1–18).

Regulation (EU) No 536/2014 of the European Parliament and of the Council of 16 April 2014 on clinical trials on medicinal products for human use, and repealing Directive 2001/20/EC (OJ L 158 of 27 May 2014, p. 1).

Verordnung über die Anwendung der Guten Herstellungspraxis bei der Herstellung von Arzneimitteln und Wirkstoffen und über die Anwendung der Guten fachlichen Praxis bei der Herstellung von Produkten menschlicher Herkunft (Arzneimittel — und Wirkstoffherstellungsverordnung — AMWHV) (Ordinance governing the Implementation of Good Manufacturing Practice in the Manufacture of Medicinal Products and Active Substances and governing the Implementation of Good Professional Practice in the Manufacture of Products of Human Origin [Medicinal Products and Active Substance Manufacturing Ordinance]) of 3 November 2006 (Federal Law Gazette I, p. 2523), last amended by Article 3 of the Regulation of 11 February 2013 (Federal Law Gazette I, p. 188).

Verordnung über die Anforderungen an Qualität und Sicherheit der Entnahme von Geweben und deren Übertragung nach dem Transplantationsgesetz (TPG-Gewebeverordnung — TPG-GewV) (Ordinance governing the requirements for quality and safety of the procurement of tissues and their transplantation according to the Transplantation Act [TPG Tissue Ordinance]) of 26 March 2008 (Federal Law Gazette I, p. 512), amended by Article 1 of the Regulation of 28 May 2014 (Federal Law Gazette I, p. 600).

Verordnung über das Register der Gewebeeinrichtungen nach dem Transplantationsgesetz (TPG-Gewebeeinrichtungen-Registerverordnung — TPG-GewRegV) (Ordinance governing the Register of Tissue Establishments according to the Transplantation Act [TPG Tissue Establishments Register Ordinance]) of 15 December 2008 (Federal Law Gazette I, p. 2446), amended by Article 2 (9) of the Act of 22 December 2011 (Federal Law Gazette I, p. 3044).

Verordnung über die Anwendung der Guten Klinischen Praxis bei der Durchführung von klinischen Prüfungen mit Arzneimitteln zur Anwendung am Menschen (GCP-Verordnung — GCP-V) (Ordinance governing the implementation of Good Clinical Practice in Performing Clinical Trials with Medicinal Products for Human Use [GCP Ordinance]) of 9 August 2004 (Federal Law Gazette I, p. 2081), last amended by Article 8 of the Act of 19 October 2012 (Federal Law Gazette I, p. 2192).

12

LEGAL ISSUES ASSOCIATED WITH PRODUCING AND MARKETING TISSUES INTERNATIONALLY

Caroline A. Hartill

OVERVIEW OF KEY CHALLENGES

Tissue banks in North America, Europe and parts of Asia (e.g., Korea) are actively distributing tissue grafts to markets beyond their national borders. The legal framework in which tissue banks must operate is complex in that both ethical concerns focused on donors and recipients, as well as regulatory requirements apply. Ethical concerns are often driven by cultural norms. Historically the laws applied to tissue donation, particularly when dealing with cadaveric donors, are modeled on those for organ donation. Subsequently this has led to a complex matrix of requirements for tissue banks to consider. The most common issues relate to the differences between opt-out systems of donation (presumed consent), where all individuals in that jurisdiction, country or region are presumed to be willing to donate unless they opt out. There is also first-person informed consent, also described as authorized donation. Both of these are accomplished via registries which the potential donor will join while still alive, and next-of-kin consent where a family member or authorized advocate is approached to consider donation. Beyond the donations aspects, there are also laws and guidelines in some geographical areas which address the "commercial" aspects of tissue banking activities. This is often simplified into a "for-profit" or "not-for-profit" model for tissue banks. Unfortunately, this definition represents an oversimplification of the models via which tissue banks have historically developed, and continue to do so. Tissue processors may operate in a tax-exempt model or in a private model for financial purposes. However, the overall business model for these banks is the same. They acquire donors in accordance with ethical and legal donation and consent norms in their local jurisdiction and then process and distribute tissue in the same commercial manner. The primary challenge for tissue banks in this respect is navigation of the laws and guidelines regarding consent and distribution models in the destination country or region. In most instances the laws regarding consent in the country of graft origin are accepted as a basis for their distribution in other countries even though the related laws may be different in that jurisdiction. On the other hand, the question of commercialization remains a challenge. Currently there is a lack of clear understanding about the reality of tissue bank business models which needs to be addressed, to ensure that ethical practices are defined and clearly understood across geographical boundaries.

Next we consider the array of regulations for tissue banks. There are highly developed regulations in some areas of the world, whereas in other parts there are no regulations at all. There is no regulation to address importation of tissue from another country, or a default to applying medical device type regulation to tissue implants. This is further complicated by the fact that in some countries certain tissue implants, such as demineralized bone matrix (DBM) for example, when mixed with a material such as glycerol or gelatin (non-human material) in order to allow proper placement or dispensing of the DBM, are then regulated as medical devices in their own right. Here we focus on understanding the more significant and challenging differences in regulation when making tissue implants available outside the country of origin.

In countries where there is regulation of tissue banks and tissue banking activities, it is usual to see some level of regulation related to tissue importation. This holds true for example, for the United States as well as Canada, Mexico, China, Singapore, and some European Union (EU) countries. It is important to note that within the EU there are significant differences in approach to tissue regulation when compared to other medical products and implants such as medical devices. At this time tissue implants (of human origin) are explicitly excluded from coverage under the Medical Device Directive (93/42EC) which provides for safety and regulation of all medical devices including implants. Several EU directives were published during the last several years, and provide a framework for regulation across the EU. The intent was to develop a common market framework to at least facilitate the movement of tissue implants within the EU among member states. Unlike the medical device directives, each member state was allowed to create specific regulations in addition to the directives. The result is that even between member states of the EU, free movement of tissue is not a reality. Examples include Germany, France, and Italy where widely differing regulations have developed, which in effect eliminate and contradict the common market approach. Tissue banks seeking to distribute tissue across jurisdictions have developed sophisticated regulatory measures and distribution controls to ensure local (destination) requirements are met. Such models include development of regulation comparison matrices and adherence to voluntary standards such as those produced by the American Association of Tissue Banks (AATB), for example, *Standards for Tissue Banking*, currently in its 13th edition, containing accreditation policies and guidance documents, which lead to accreditation by the AATB. AATB has successfully promulgated these standards beyond the United States and has accredited several tissue banks in other countries, such as Korea, Singapore and Canada. Both of these approaches are helpful in ensuring that a broad range of regulations are met as well as identifying specific differences.

DEVELOPMENT OF ADVANCED TECHNOLOGIES FOR TISSUE PROCESSING

Of particular note is that, from a regulatory perspective, there is no consistent requirement to sterilize human tissue for implantation. Rather, the emphasis of regulation is on prevention of contamination and cross-contamination during tissue procurement and processing. This holds true for the United States, where the tissue regulation paradigm 21CFR 1271, Human Cells, Tissue and Cellular and Tissue-Based Products, promulgated from 2001 to 2005 does not require disinfection or sterilization of tissue intended for transplant but does require that if a claim of bio-burden reduction or sterility assurance is made, it must be based on appropriate validated methods. This is true in regulations for many other parts of the world, for example the EU directives. These can provide an overarching regulatory framework for all EU member states, such that the same principle applies throughout the EU. Nevertheless, certain EU member states such as Germany have implemented additional specific requirements related to tissue safety and viral inactivation, as described in Chapter 11.

Regulations are often developed in a reactionary manner This is especially true for tissue banks since the origin of most regulation in this area was developed to address the risk of transmission of blood-borne pathogens. In particular, viruses have been known to be transmitted by blood transfusion and in the 1980s there were incidences of HIV transmission by organs and tissues prior to the availability of testing donors. Often regional differences were encountered in donor screening and testing criteria. These differences present challenges to the distribution of tissue grafts to patients in need of a particular graft or type of graft across national borders. Examples of this include the US-based criteria related to the risk of transmission of variant Creutzfeld-Jacob disease (vCJD) based on geographical location of the donors along with their historical travel pattern. This particular requirement alone eliminates the possibility for importing tissue transplants into the United States from most of the European continent and also grafts from American citizens who have traveled or lived extensively in Europe. This criterion is not applied under European regulation and there is no recorded transmission from a tissue transplant. Even within Europe there are differences in requirements for the type of blood tests to be performed. In France, for example, additional types of testing for HIV are required, which is a legacy of the challenges faced historically in France in the area of blood transfusion.

Currently most regulation of tissue banks and tissue processing is based on the following:

- Tissue donors are heavily screened according to criteria similar to those applied to screening of blood donors;
- A blood sample is tested for certain viral markers (actual testing requirements may vary);

- The concept that at the point of recovery, tissue is not inherently potentially contaminated;
- That providing proper surgical and aseptic techniques are deployed during the actual recovery of tissue, along with processing controls and conditions to prevent cross-contamination, the likelihood of viral disease or bacterial contaminant transmission is slight.

We know this to be potentially a flawed concept. While the risk of donor tissue-associated contamination causing adverse outcomes in a recipient is small, it does exist. Cases of tissue-related disease transmission have been reported for bacteria, in one notable case in the United States in 2001 leading to the death of a young person who received a tissue transplant in knee surgery, as reported in the Mortality and Morbidity Weekly Report (MMWR) published by the USA Centers for Disease Control (CDC), (MMWR March 15, 2002, **51**(10), 207–210). More recently, in 2011 a viral disease transmission was reported in a young child, when hepatitis C was transmitted (MMWR December 23, 2011, **60**(50), 1697–1700).

In the United States, since the late 1990s some tissue banks (processors) have focused efforts on reducing the potential for allograft-related disease transmission. Proprietary methods have evolved and are routinely used. These are proved to reduce, eliminate, or destroy bacterial contaminants (microorganisms and spores). Washing and other mechanical and chemical means can reduce and/or eliminate viral components if present. Some banks have also validated processes that can inactivate viruses (Kainer *et al.*, 2004). While these developments potentially enhance tissue safety and reduce risk to patients, a varied and diverse approach has been taken toward these efforts and results vary accordingly. A factor that provides an additional challenge is that no regulation specifies performance requirements for a given type of tissue graft or implant. As disinfection and sterilization methods are developed, often tissue processors have implemented measures to increase tissue safety with limited (or no) studies to determine the potential effect on clinical utility of the graft after treatment by these methods. Examples include the introduction of ethylene oxide and of irradiation techniques for sterilization of some tissue such as tendons and ligaments often used in treatment of knee injuries involving the repair of ligaments. Such techniques are highly effective in addressing the risk of bacterial contamination of the tissues. However, the literature illustrates that the use of ethylene oxide was fairly quickly recognized to cause an acute inflammatory reaction in the patient and was soon discontinued as a processing method. For the treatment of these types of tissue with irradiation, there is no indication of an acute affect. However, there is currently a growing body of literature (Pan *et al.*, 2009; Guo *et al.*, 2012; Sun *et al.*, 2009; Rappe *et al.*, 2007) indicating that this processing method may be detrimental to the tissues in terms of their clinical utility. On the other hand, many tissues such as bone can be effectively treated with irradiation with limited or no effect on the clinical utility of the graft. The same principle applies to other processing methods where certain preservation, disinfection, or sterilization techniques are suitable for certain tissue based on the needs of clinical performance, and may not be suitable for the tissue for all clinical applications.

Donor screening and processing methodology can have a significant positive impact on reducing risk of disease transmission. Clinical utility should also be considered when applying these processing methods, and where preservation of key attributes of the tissue is essential to good clinical outcomes.

Given the relative inconsistency across international regulations, the advances of many processing methodologies beyond the regulatory requirement, coupled with various types of validation and testing performed by processors, it is often difficult, if not impossible, for surgeons, hospitals, patients and their counsel to readily ascertain the actual or potential relationship between an adverse event or poor clinical outcome, and the graft utilized. This situation also provides a challenge to tissue banks as questions arise from a variety of sources over time regarding regulation, patient outcomes, and risk. More harmonization of regulation is desirable but unlikely to occur quickly. Performance standards and methods to measure tissue safety and utility need to be developed. This will help to ensure consistency and transparency for all stakeholders.

Information related to a particular tissue or processing method is often provided to hospitals and surgeons but in many cases, not clearly understood or indeed shared with patients. Clear information about residual risk/benefit should be freely available to stakeholders. Availability of this information is also necessary to alleviate concerns among stakeholders and reduce the liability of tissue banks.

REFERENCES

Archibald LK, Jernigan DB, Div of Healthcare Quality Promotion, National Center for Infectious Diseases; Kainer MA, EIS Officer, CDC (2002) Allograft-Associated Bacterial Infections, United States; March 15, 2002, **51**(10); 207–210.

Guo L *et al.* (2012) Anterior cruciate ligament reconstruction with bone-patellar tendon-bone graft: Comparison of autograft, fresh-frozen allograft, and γ-irradiated allograft, *Arthroscopy* **28**(2), 211–217.

Kainer MA, Linden JV, Whaley DN, Holmes HT, Jarvis WR, Jernigan DB, Archibald LK (2004) Clostridium infections associated with musculoskeletal-tissue allografts, *N Engl J Med* **350**(25), 2564–2571.

Marvin MR, Green SK, Thoroughman D, Sugg TJ, Humbaugh KE, Vaz LE, Burchett SK, Moffitt K, Nielsen CF, Holmberg SD, Drobeniuc J, Khudyakov Y, Kuehnert MJ, Hocevar SN, Mahajan R (2011) Transmission of Hepatitis C Virus through Transplanted Organs and Tissue, Kentucky and Massachusetts; December 23, 2011, **60**(50); 1697–1700.

Pan PC, Liu JZ, Sun K (2009) Biomechanics of gamma-irradiated versus non-irradiated human tendon. *Pittsburgh Orthopaedic Journal.*

Rappe M, Horodyski M, Meister K, Indelicato PA (2007) Non-irradiated versus irradiated achilles allograft: *In vivo* failure comparison, *American Journal of Sports Medicine* **35**(10), 1653–1658.

Sun K, Tian SQ, Zhang JH, Xia CS, Zhang CL, Yu TB (2009) Anterior cruciate ligament reconstruction with BPTB autograft, irradiated versus non-irradiated allograft: A prospective randomized clinical study, *Knee Surgery Sports Traumatology and Arthroscopy* **17**, 464–474.

13

A RETROSPECTIVE LOOK AT THE LARGEST RECALL OF PROCESSED BONE TISSUE IN HISTORY

Helena Lazares and John T. Salvucci

> Human corpses "sat in an alley, unrefrigerated for days, and, after tissue was removed, all that would remain was a head and bloody torso."
>
> *Com. v. Mastromarino*, 2 A.3d 581, 588 (Pa. Super. Ct. 2010)

INTRODUCTION

Beginning on September 30, 2005, a series of announcements was made by bone tissue processors concerning discrepancies in donor documentation provided by Biomedical Tissue Services Ltd (BTS). Through October 14, 2005, the processors each announced a voluntary market withdrawal and/or recall of allograft derived from BTS donors. The Food and Drug Administration (FDA), on October 26, 2005, released an announcement that included concerns that BTS may not have adequately screened donors for infectious disease. The FDA ordered BTS operations to be shut down. BTS had supplied tissue for several years to at least five companies that processed the tissue into allografts, including Regeneration Technologies, Inc., Tutogen, LifeCell Corporation, Central Texas Regional Blood and Tissue Center, and Lost Mountain Tissue Bank, Inc. All of the bone tissue processed into allografts from tissue recovered by BTS were the subject of an October 2005 FDA supervised recall. The FDA and Centers for Disease Control and Prevention (CDC) recommended that implanting physicians offer to provide patients access to testing for the diseases for which donors were generally screened and report any positive test results to the processors and/or the FDA directly. From June 2002 through October 2005, approximately 25,000 BTS-recovered tissue products were distributed to all 50 states and internationally.

In May 2006, the CDC in its morbidity and mortality weekly report discussed its investigation into the BTS recalled tissue. In that report the CDC repeated its recommendation that allograft recipients be tested for hepatitis B (HBV), hepatitis C (HCV), human immunodeficiency virus (HIV), and syphilis. Plaintiffs in the United States who underwent surgery in which one or more allografts from BTS were implanted initiated lawsuits against funeral homes, recovery agencies, processors, and a distributor. The first lawsuit was filed on November 18, 2005, and as of October 2007 there were over 700 plaintiffs suing in US state and federal courts, that total eventually reaching 1,000 claims.

The principals of BTS were indicted for crimes related to the unauthorized recovery of tissue and falsification of donor screening records. Prison sentences were issued to numerous parties involved, both for unauthorized recovery and for failure to meet the aseptic processes dictated by best practices.

After litigation commenced, the cases were consolidated and assigned to one federal judge for pre-trial proceedings on June 21, 2006. The defendants jointly filed a motion for summary judgment on October 16, 2007. On September 4, 2008, oral argument was conducted before the Federal Court, and thereafter on October 22, 2008, the Federal Judge issued a 105-page opinion in which he concluded that the bone tissue implanted was not capable of transmitting HBV, HCV, HIV, syphilis, or cancer.

Fortunately, to this day there has been no report of disease transmission by bone tissue recovered by BTS and distributed by any of the processors. The failure by the CDC and the FDA to announce any case of disease transmission, now, nine years later, adds to the weight of the scientific conclusion that disease transmission could not occur.

THE CRIMINAL SCHEME THAT INITIATED THE RECALL

At the beginning of the twenty-first century, the demand for human tissue far surpassed available supply; a man named Michael Mastromarino decided to capitalize on this shortage. In 2002, Mastromarino created a corporation called Biomedical Tissue Services (BTS), through which he recovered human tissue from cadavers and supplied them to tissue banks around the United States. Upon founding BTS, Mastromarino, a former dentist whose license was suspended due to drug abuse, became licensed as a recovery agency for human tissue and began seeking out funeral homes to provide him with access to cadavers.

Mastromarino originally worked with a few funeral homes in New York and New Jersey. These homes provided him with access to cadavers from which he and his team of "cutters" could harvest tissue without the consent of the deceased or their next-of-kin. Mastromarino reportedly paid undertakers up to $1,000[1] per corpse.

Mastromarino and his team reconstructed the cadavers with PVC pipe to conceal the unapproved removal of tissue. Thereafter he sought a simpler process to further his criminal enterprise. An arrangement with a funeral director who had access to a crematorium would provide him with cadavers that would not need to be reconstructed after harvesting of the tissue was complete. Mastromarino met with the co-owners of Liberty Crematory to foster this arrangement, which led to a significant increase in his recovery of bone tissue.

In February 2004, Mastromarino began working with James McCafferty, Louis Garzone, and Gerald Garzone. Each operated separate funeral homes and were partners in a crematorium business, Liberty Cremation, Inc., in Philadelphia, Pennsylvania. Together, Mastromarino, the Garzones, and McCafferty, worked to illegally enrich themselves by trafficking in stolen body parts. Without obtaining the families' consent, they procured bones, skin, tendons, spines, and other tissue from the dead entrusted to the care of Philadelphia funeral homes for cremation and burial.

The harvested tissue was taken from corpses with flagrant disregard for regulations and procedures designed to ensure the safety of body parts. Safe tissue protocols require tissue removal within 15 hours of death. The bodies Mastromarino and his team harvested tissue from routinely sat for days without refrigeration, sometimes in an alley or a garage. One corpse reportedly sat for 113 hours after death, another for 100 hours.

[1] All monetary references are in US dollars.

The cutters used power tools to remove spines and bones from the bodies. Without the need to prepare the body for a viewing, Mastromarino and his team went "whole hog" on the bodies to be cremated. The process of harvesting tissue by regulatory standards would take several hours, yet Mastromarino and his "cutters" tore bodies apart in 30 minutes taking everything they could.

The lack of respect for the deceased, illegality of harvesting, and deliberate neglect of regulatory law were not the end of the scheme. In order to sell the tissue they harvested, Mastromarino and his associate, Lee Cruceta, falsified records and forged documents. Mastromarino and Cruceta falsified facts including name, time of death, and health condition of the cadavers. Mastromarino deliberately replaced blood samples from bodies that might have been infected with HIV and HCV, evading tests designed to screen for these diseases, and fabricated medical histories for individuals who died from cancer or septic diseases.

Mastromarino and his team took the harvested tissue from the diseased and decaying bodies, and sold it to companies that then distributed bone tissue to hospitals worldwide for use in medical procedures. BTS supplied tissue for several years to at least five companies that processed the tissue into allografts. Overall, the total amount tissue processing companies paid to BTS for tissue stolen in all of the funeral homes was approximately $5,000,000.

Several companies distributed for implantation thousands of pieces of BTS tissue, some of it from infected corpses, all of it dangerously mishandled and mislabeled. One company alone sold more than 2,000 pieces of tissue stolen from bodies at the Philadelphia funeral homes. At least five Philadelphia-area and 41 Pennsylvania hospitals implanted BTS tissue in patients.

New York Detective Patricia O'Brien sparked the investigation that uncovered the scheme when she investigated a Brooklyn funeral home for stolen funds in the fall of 2004. When she visited the home, she noticed a room that looked like an operating room. She looked into the home's files and discovered that the previous owner had been running more than a funeral business.

Mastromarino, Cruceta, and the funeral directors had conducted their operation for three-and-a-half years under the noses of federal and state regulators and private auditors before they were caught. The Philadelphia grand jury report concluded that the crimes "went undetected because current regulations and overseers failed to take into account the enormous incentives and opportunities to commit theft and fraud in the body tissue industry." (In re County Investigating Grand Jury XXI, 2007).

After the ghoulish scheme was uncovered, BTS records showed that they had procured tissue from a total of 1,077 bodies in three states: Pennsylvania, New Jersey, and New York.

In 2006, the case was submitted for review to the Grand Jury. The Grand Jury Report recommended that officials prosecute Mastromarino and his accomplices to the fullest extent possible. Mastromarino pled guilty to 1,353 separate counts including corrupt organization, conspiracy, theft by unlawful taking, deceptive business practices, and abuse of corpse. He was sentenced on October 22, 2008 to an aggregate of 25 to 58 years in prison.

Louis and Gerald Garzone were each sentenced to eight to 20 years in prison for illegally selling body parts for use in surgeries. Lee Cruceta was also sentenced to eight to 20 years for his part in the criminal scheme. James McCafferty was sentenced to three-and-a-half to ten years.

In an ironic twist of fate, Michael Mastromarino died in July 2013, while serving out his prison sentence, from complications, including bone cancer.

FEDERAL AND LEGAL INTERVENTION

Nearly a year after Detective O'Brien's suspicions sparked the investigation into Mastromarino's scheme, the FDA and CDC became involved. On September 29, 2005, a tissue-processing company discovered inaccuracies in donor records and notified the FDA. This launched an FDA investigation into BTS, which determined that tissues recovered from BTS might not have met donor eligibility requirements and were not properly screened for certain infectious diseases.

The FDA Current Good Tissue Practice rules require manufacturers to recover, process, store, label, package, and distribute human tissue products to prevent introduction, transmission, or spread of communicable diseases. FDA regulations require that each donor be screened and tested prior to accepting their tissue. Donors are screened through a review of their medical records, a physical assessment, and questioning of their next-of-kin to determine whether clinical evidence of relevant communicable diseases exists. Blood samples from each donor must be provided to the testing laboratory for donor testing of HIV, hepatitis B, hepatitis C, and syphilis before tissues are released for distribution.

The FDA investigation determined that information regarding cause of death for some BTS donors was not consistent with death certificate data obtained from the states where the deaths occurred. The FDA also determined that BTS had failed to control conditions adequately during tissue recovery and did not recover tissues in a manner that would prevent contamination.

BTS supplied tissue over several years to at least five companies that processed the tissue into allografts. Most of these tissue allografts were bone or demineralized bone matrix; others included skin and soft tissue. The tissues were sent to processors, who distributed them through one or more sub-distributors or directly to clinicians and healthcare facilities. Before distribution, tissues were disinfected by tissue processors to reduce or eliminate contamination with bacteria, fungi, or viruses. Over 10,000 allografts of recovered BTS tissue were implanted.

In October 2005, the FDA supervised a recall of all tissue recovered by BTS. The five processing companies that received BTS-recovered tissue products, Regeneration Technologies, Inc., Tutogen, LifeCell Corporation, Central Texas Regional Blood and Tissue Center, and Lost Mountain Tissue Bank, Inc., voluntarily recalled all tissue. The processing companies sent letters to healthcare facilities notifying them of the recall and recommending they provide access to testing for transplant recipients.

About two weeks later, the FDA notified the public through an announcement providing information about its investigation into human tissue recovered by BTS (FDA News Release, 2005). The release stated that the FDA and CDC believed the risks of contracting infectious disease from these tissues were low because the tissues were routinely processed using methods that reduce the risk of infectious disease, and that no adverse reactions related to these tissues had been reported to the FDA at that time. The release, however, recommended "implanting physicians inform their patients that they may have received tissue from a donor for whom an adequate donor eligibility determination was not performed." (FDA News Release, 2005). The release further noted that while the overall risk of infection was likely low, patients should be provided access to testing for the relevant communicable diseases: HBV, HCV, HIV, and syphilis.

The first civil lawsuit brought by an allograft recipient was filed in New Jersey Superior Court, Middlesex County, on November 18, 2005, and the case was removed to a federal court. After that, over 700 plaintiffs filed lawsuits in federal and state courts against funeral homes, recovery agencies, processors, and a distributor. Plaintiffs were patients who received allograft implants. On June 21, 2006, the Judicial Panel on Multidistrict Litigation consolidated the cases filed in federal court as *In re Human Tissue Products Liability Litigation* in the United States District Court for the District of New Jersey.

In March 2006, the FDA issued an update to its October 2005 press release (FDA News Release, 2006). This update reported that the FDA had additional information and confirmed that blood samples did not come from the same donor as the linked tissues in some instances. The FDA and CDC again recommended that physicians offer patients access to testing for HBV, HCV, HIV, and syphilis. The update also noted that patients should report any adverse reaction possibly related to a tissue transplant to their tissue processing firm.

In its May 26, 2006, Morbidity and Mortality Weekly Report, the CDC reported on its investigation with the FDA into the BTS recall (CDC MMWR, 2006). In the report, the CDC recommended testing for patients who had their allografts implanted more than six months previously. The CDC advised that patients who tested negative for disease after implants in place for six months did not need further testing.

On October 20, 2006, defendants in the consolidated federal litigation, processors and distributors of allografts, Medtronic, Inc., Medtronic Sofamor Danel USA, Inc., SpinalGraft Technologies, LLC, and Regeneration Technologies, Inc., filed a motion to dismiss plaintiffs' complaints for failure to state a claim. Defendants argued that even if

BTS failed to properly screen donors to minimize the risk of disease transmission, no virus could survive the sterilization process and storage at room temperature for a prolonged period of time.

The Court held oral arguments on March 28, 2007. At that time, the Court reserved on the motion to dismiss and converted it into a summary judgment motion on the issue of causation. The same defendants jointly filed a motion for summary judgment on October 16, 2007 on the issue of general causation for the various tort claims asserted by the recipient plaintiffs.

The motion centered on questions of science previously posed by the court. The court asked the parties to assume that the bone tissue from the cadavers was infected with HCV, HIV, syphilis, cancer, and prions, and had not undergone any processing. The court then posed two questions: (1) Can this bone tissue still transmit one of these diseases to a donee?; (2) What is the incubation period for these diseases? These questions had been suggested by the defendants, since it allowed the court to bypass any need to assess the merits of the complicated sterilization processes. Consequently, the sterilization procedures would be additional steps in ensuring that the bone allografts do not infect recipients with infectious diseases.

At oral argument, the defendants argued that the admissible evidence in this case established that there was no risk of disease transmission. In answering the first question posed by the court, the defendants maintained that lyophilized bone tissue stored at room temperature for 30 days or more cannot transmit HBV, HCV, HIV, syphilis, or cancer. The defendants further argued that the incubation period for each of the diseases in question is six months or less, and all of the plaintiffs received their allograft implants more than six months ago.

Ultimately, in a 105-page opinion, Federal Judge Martini granted summary judgment for the defendants excluding some of plaintiffs' expert testimony and ruling in favor of the defendants' arguments on the two proposed questions.

DISEASE TRANSFERS BY BONE

Judge Martini was able to rule in favor of the defendants' motion for summary judgment in the BTS litigation based on their expert testimony and related literature review. Through their memorandum of law and oral argument, the defense presented a reliable scientific history of the capability of disease transfers by lyophilized bone tissue. The plaintiffs failed to present reliable scientific and medical evidence to counter these arguments. Rather, they only offered subjective belief and unsupported speculation.

Over the past 65 years bone tissue allografts have resulted in 17 cases of viral disease transmission, two cases of bacterial disease transmission, and one case of transmission of cancer.

Bone Tissue Transmitting Viral Disease

Lyophilized bone tissue has been in use since 1951. There are no known cases of HIV or disease transmission through lyophilized bone. Only frozen bone tissue allografts, not lyophilized bone tissue allografts, have caused infection. There have been 17 cases of viral disease transmission.

Shutkin reported the first case of the transmission of viral disease by frozen bone in 1954 (Shutkin, 1954). This case concerned a medical student who became infected with HBV transmitted from donor bone. Subsequently there have been nine reported cases of HIV transmission and 7 cases of HCV transmission, all of which resulted from implantation of frozen bone allografts. (Table 13.1 details the reported cases of recipients infected with HIV, HCV, or HBV.)

Freezing preserves the infectivity of enveloped viruses, presumably by keeping their lipid membranes intact for years (Salvucci, 2011). In order for viruses to replicate, a fusion event must occur between the virus's lipid envelope and the cell's lipid membrane. This fusion event results in the building of a bridge enabling the viral DNA or viral RNA to enter the cell (Tscherne *et al.,* 2006). Water is an integral part of the maintenance of the virus's lipid envelope. Lacking water, the lipid membrane will collapse from a lamellar bilayer shell into hexagonal arrays and sheets (Bode and Read, 2000).

When moisture is removed from the bone through drying or lyophilization, the enveloped viruses are inactivated by destroying the lipid envelope. Several studies have shown that viral infectivity rapidly dissipates following drying and storage at room temperature. Laboratory studies have shown that HIV survives when dried at room temperature for at most seven days (Sattar and Springthorpe, 1991), and HBV, dried

Table 13.1. Reported cases of recipients infected with HIV, HCV, and HBV.

Year disease transmitted	Year reported	Allograft	Processed/Unprocessed	Number and disease	Reference
1954	1954	Cancellous bone probably from femoral head	Frozen/Unprocessed	1 HBV	Shutkin
1984	1988	Frozen/femoral head	Unprocessed	1 HIV	CDC-MMWR
1984–1985	1996	Bone	Frozen/Unprocessed	4 HIV	Schratt
1985	1991 and 1992	2 femoral heads and 1 tendon with bone	Frozen/Unprocessed	3 HIV	Simonds
1990	1992	Femoral head	Frozen/Unprocessed	1 HCV	Eggen
1990	1995	1 patellar tendon and 1 femoral head	Tendon in bacitracin and polymyxin and frozen; and femoral head was placed directly in freezer	2 HCV	Conrad
1996	2001	Femoral head	Unspecified, but assumed frozen based upon placement into hospital bone bank; also unprocessed	1 HIV	Chien-Min Li
1998	2003	Cancellous bone chips	Soaked in antibiotic, alcohol, and frozen	1 HCV	Trotter
2000	2005	Three tendons with bone	Lavaged with sterile water, then soaked in isopropyl alcohol, antibiotics, and fresh frozen	3 HCV	Tugwell

on a stainless steel surface, can survive for a period of up to 14 days (Sattar and Springthorpe, 1996).

In a 2007 study, chimpanzees were injected with three HCV infectious aliquots, which were previously air dried at room temperature for 16 hours, four days, and seven days (Kamili *et al.*, 2007). The Kamili study reported that the aliquot dried for 16 hours transmitted HVC, while the aliquot dried for four days did not.

Studies specifically dealing with lyophilization have shown that enveloped viruses survive lyophilization, yet they establish a rapid loss of infectivity on storage. In a 2005 study, the enveloped virus, vesicular stomatitis virus (VSV), was tested following lyophilization and a three-to-four-hour log reduction was noted (Uhlenhaut *et al.*, 2005). The time elapsed between lyophilization and testing of VSV was less than one hour. In a similar study, feline leukemia virus (FeLV) was determined to survive lyophilization, but for a period of less than seven days (Crawford *et al.*, 2004).

The life of the lipid envelope can be extended if it is protected by freezing, or by adding cryopreservative agents prior to drying (Salvucci, 2011). Drying without cryopreservative agents accelerates the progression toward inactivity. Air-drying and lyophilization do not inactivate enveloped viruses; however, the process of water removal using either method limits infectivity of the viruses to a certain period of time. Consequently, the evidence leads to the conclusion that bone, lyophilized and stored at room temperature for 15 days or more, is not capable of transmitting HIV, HBV, and HCV.

Bone Tissue Transmitting Bacterial Disease

Disease transmission resulting in bacterial infection would be known immediately following implantation of tissue. Although there have been two instances of bacterial disease transmitted by allografts (summarized in Table 13.2), there were no claims for bacterial disease involved in the BTS litigation.

Bone Tissue Transmitting Cancer

There is only one reported case of cancer transmitted by frozen bone allografts (see Table 13.3).

Table 13.2. Reported cases of recipients infected with bacterial disease.

Year disease transmitted	Year reported	Allograft	Processed/ Unprocessed	Number and disease	Reference
1953		Ribs	Frozen and solution containing penicillin and streptomycin	1 tuberculosis	James
2001	2001	Knee osteochondral (bone-cartilage)	Processed	1 clostridium	CDC-MMWR

Table 13.3. Reported case of recipient infected with cancer.

Year disease transmitted	Year reported	Allograft	Processed/ Unprocessed	Number and disease	Reference
1991	1997	Femoral head	Fresh frozen unprocessed	1 HTLV-1	Sanzen

TRANSMISSION OF VIRAL DISEASE IN BTS LITIGATION

None of the represented plaintiffs in the BTS litigation received frozen bone tissue allografts. Based on the plaintiffs that provided tissue identification information, the defendants determined that the implanted allografts had been lyophilized and stored at room temperature for a period of over 30 days in all but one instance. In the outlying instance, the allograft was stored for 22 days at room temperature and that plaintiff tested negative for infectious disease ten months after implantation surgery.

All parties agreed that frozen bone allografts could transmit disease. Additionally, all parties agreed that unprocessed bone tissue stored at room temperature could transmit HIV, HBV, HCV, syphilis, and cancer for an initial window of time. The court focused its inquiry at oral argument on the period of time lyophilized bone tissue allografts stored at room temperature can transmit disease; and, the incubation period for detection.

In answering the two questions posed by the court, the defendants maintained that bone tissue stored at room temperature for 30 days or more cannot transmit HBV, HCV, HIV, syphilis, or cancer. The defendants further argued that the incubation period for each of the diseases in question is six months or less, and all of the represented plaintiffs received their allograft implants more than six months previously. Plaintiffs argued for no specific period of time in which transmission can occur, but that it is longer than 30 days. Plaintiffs further argued that the detection of HBV and HCV infection could take up to two years and HIV infection between six and eight months.

Plaintiffs and defendants each presented expert opinions on the time period of infectivity for bone tissue stored at room temperature. Experts on both sides agreed that viruses lose infectivity over time when kept at room temperature. Plaintiffs' experts, Dr Busch, a blood specialist, and Dr Kowalski, a microbiologist, acknowledged that enveloped viruses, like HBV, HCV, and HIV, lose infectivity at room temperature over time. Plaintiffs' expert Dr Busch testified that he did not recall "in those experimental studies a time period beyond 14 days referenced with respect to residual infectivity" (In re Human Tissue Products Liab. Litig., 2008).

Plaintiffs' expert Dr Parisian was unable to explain why the studies she relied on supported her conclusion regarding the ability of bone allografts to transmit disease beyond 30 days. She further based her determination that there is always a risk of disease transmission without donor screening because of the CDC's reliance upon both donor screening and processing to ensure the safety of bone allografts. The court found her opinions, at best, nothing more than pure speculation.

The court additionally found that Dr Kowalski's opinions suffered from the same issues as Dr Parisian's opinions in that he failed to articulate the scientific and medical rationale for extrapolating the results of the available studies to the set of facts in the litigation.

Defendants presented studies by Van Bueren and Resnick, which showed substantial reductions of HIV titer over 30 days or less, even with the use of stabilizers to protect the virus. Defendants' experts on HIV, Drs Kuritzkes, Richman, Rutala, Kainer, and Jarvis, all agreed that bone tissue kept at room temperature for 30 days or more cannot transmit HIV.

Neither plaintiffs' nor defendants' experts opined that syphilis can be transmitted by unprocessed bone tissue. Syphilis is a bacterial disease and can survive only briefly in the environment outside of blood. In whole blood at temperatures much lower than room temperature, syphilis loses infectivity in, at most, five days. Similarly, plaintiffs failed to provide any argument to suggest bone tissue can transmit cancer after 30 days or more. Plaintiffs additionally failed to provide more than speculative opinions regarding the risk of transmitting prion disease.

After an in-depth review of plaintiffs' expert opinions, Judge Martini concluded that plaintiffs failed to establish credible expert testimony to conclude that lyophilized bone tissue stored at room temperature could transmit any of the named diseases after 30 days.

The court next addressed the issue of incubation. The standard adopted by the CDC and FDA is that the incubation period for HBV, HCV, and HIV is less than six months.

Without providing any scientific or medical articles to support his opinions, plaintiffs' expert Dr Manzarbeitia offered three different opinions with respect to the incubation period for HCV, including "up to twelve months," "six weeks to ten years" and "six weeks and eighteen months to two years with appropriate testing" (In re Human Tissue Products Liab. Litig., 2008). Plaintiffs' expert Dr Klein opined that the incubation period for HBV and HCV was "two years or greater" from the time when a person has been exposed to when that person will test positive (In re Human Tissue Products Liab. Litig., 2008). Dr Klein based his opinion upon his clinical experience in hepatology and studies by Beld, Prieto, and Dickson, which all concluded that the period from infection to detection via antibody testing for HBV in the liver transplant population and HCV among injecting drug users was extended to over two years.

Plaintiffs' experts Manzarbeitia and Klein relied heavily on extrapolating experience with liver transplants. The court found the extrapolation of experience with organ transplants to bone allograft transplants unreliable, as each expert failed to provide any basis for his extrapolation. Indeed, it would be hard to equate the transplanting of live blood filled organs with the transplanting of lyophilized bone stored at room temperature in excess of 30 days.

Defendants moved to exclude the proposed testimonies of Drs Parisian, Kowalski, Klein, and Manzarbeitia regarding the transmission of HBV, HCV, HIV, syphilis, cancer, and prion disease, and the proposed testimonies of Drs Klein and Manzarbeitia regarding the length of the incubation period for failure to meet the standards for the admission of expert evidence.

Judge Martini ruled to exclude the testimony of plaintiffs' experts, noting that "plaintiffs have failed to provide any reliable evidence to support their claims of general causation with respect to the transmission of HIV, HBV, HCV, cancer, and syphilis through unprocessed (lyophilized) human bone tissue that has been stored at room temperature for thirty days or more before transplantation into an individual and the transmission of prion disease through human cadaveric bone tissue." (In re Human Tissue Products Liab. Litig., 2008).

Judge Martini further held that "[w]ith the Court's exclusion of certain opinions of plaintiffs' experts, . . . the Court is left with no evidence in support of plaintiffs' claim of general causation" (In re Human Tissue Products Liab. Litig., 2008). Thus, Judge Martini granted summary judgment in favor of defendants based on the following: "(1) Unprocessed bone tissue and bone paste stored at room temperature for thirty (30) days or more is not capable of transmitting HBV, HCV, HIV, syphilis, or cancer; (2) Unprocessed bone tissue and bone paste is not capable of transmitting prion disease; (3) Federal plaintiffs who have tested negative for HBV, HCV, HIV, and syphilis more than six (6) months after their bone tissue or bone paste transplant surgery cannot establish general causation with respect to HBV, HCV, HIV, and syphilis." (In re Human Tissue Products Liab. Litig., 2008).

CONCLUSIONS

The BTS litigation highlights the importance of each tier in a system to properly cover for all risks. The tissue processing companies in this case were victimized by BTS's blatant disregard for safe tissue practices. Nevertheless, their sterilization methods, along with lyophilization and storing at room temperature for a period longer than 30 days prior to implantation, protected the public. Fortunately, disease has not been transmitted by BTS bone tissue. Indeed, to this day, neither the FDA nor the CDC has announced even a single case of disease transmission related to the recall of BTS tissue.

REFERENCES

Bode AP, Read MS (2000) Lyophilized platelets for transfusion, in Seghatchan J, Snyder EL (eds.) *Platelet Therapy; Current Status and Future Challenges*, Elsevier Science, Amsterdam, pp. 131–167.

Centers for Disease Control and Prevention (1988) Transmission of HIV through bone transplantation: Case report and public health recommendations, *MMWR* **37**(39), 597–599.

Centers for Disease Control and Prevention (2001) Public health dispatch: Update: Unexplained deaths following knee surgery, *MMWR* **50**(46), 1035–1036.

Conrad EU, Gretch DR, Obermeyer KR, Moogk MS, Sayers M, Wilson JJ, Strong DM (1995) Transmission of the hepatitis-C virus by tissue transplantation, *J Bone Joint Surg Am* **77**, 214–224.

Crawford MJ, Swenson CL, Arnoczky SP, O'Shea Jr RH (2004) Lyophilization does not inactivate infectious retrovirus in systemically infected bone and tendon allografts, *Am J Sports Med* **32**(3), 580–586.

CDC MMWR (2006) Brief Report: Investigation into Recalled Human Tissue for Transplantation, United States, 2005–2006. [Online] (Updated May 26, 2006). Available at http://www.cdc.gov/mmwr/preview/mmwrhtml/mm5520a6.htm.

Eggen BM, Nordbo SA (1992) Letter to the editor, *New Engl J Med* **26**(6), 411.

FDA news release (2005) FDA Provides Information on Investigation into Human Tissue for Transplantation. [Online] (Updated October 26, 2005). Available at http://www.fda.gov/NewsEvents/Newsroom/PressAnnouncements/2005/ucm108506.htm.

FDA news release (2006) FDA Public Health Notification: Update of Information about BioMedical Tissue Services. [Online] (Updated March 2006). Available at http://www.fda.gov/biologicsbloodvaccines/safetyavailability/ucm154999.htm.

In re County Investigating Grand Jury XXI, Misc. No. 05-011487, (Pa. D & C 2007).

In re Human Tissue Products Liab. Litig., 582 F. Supp. 2d 644, 670 (D.N.J. 2008).

James JIP (1953) Tuberculosis transmitted by banked bone, *J Bone Joint Surg* **35-B**, 578.

Kamili S, Krawczynski K, McCaustland K, Li X, Alter M (2007) Infectivity of hepatitis C virus in plasma after drying and storing at room temperature, *Infect Control Hsp Epidemiol* **28**(5), 519–524.

Li C-M, Ho Y-R, Liu Y-C (2001) Transmission of human Immunodeficiency virus through bone transplantation: A case report, *J Formos Med Assoc* **100**(5), 350–351.

Memorandum of Law in Support of Defendants' Motion for Summary Judgment at 4, In re Human Tissue Products Liab. Litig., 582 F. Supp. 2d 644 (D.N.J. 2008) (No. 2:06-CV-00135), 2007 WL 4378089, at *1.

Salvucci, JT (2011) Bone tissue, lyophilized and stored at room temperature for 15 days or more, is not capable of transmitting HIV, HCV or HBV, *Cell Tissue Bank* **12**, 99–104.

Sanzen L, Carlsson A (1997) Transmission of human T-cell lymphotronic virus type 1 by a deep-frozen bone allograft, *Acta Orthop Scand* **68**(1), 70–76.

Sattar SA, Springthorpe VS (1991) Survival and disinfectant inactivation of the human immunodeficiency virus: A critical review, *Rev Inf Dis* **13**(3), 430–447.

Sattar SA, Springthorpe VS (1996) Transmission of viral infections through animate and inanimate surfaces and infection control through chemical disinfection, in Hurst CJ (ed.) *Modeling Disease Transmission and Its Prevention by Disinfection*, Cambridge University Press, Cambridge, England, pp. 224–257.

Schratt HE, Regel D, Kiesewetter B, Tscherne H (1996) HIV-Infektion durch kaltekonservierte Knochentransplantate, *Unfallchirurg* **99**(9), 679–684.

Shutkin NM (1954) Homologous-serum hepatitis following the use of refrigerated bone-bank bone, *J Bone Joint Surg Am* **36**, 160–162.

Simonds RJ, Holmberg SD, Hurwitz RL, Coleman TR, Bottenfield S, Conley LI, Kohlenberg SH, Castro KG, Dahan BA, Schable CA, Rayfield MA, Rogers MF (1992) Transmission of human immunodeficiency virus type 1 from a seronegative organ and tissue donor. *New Engl J Med* **326**(11), 726–732.

Transcript of oral argument, In re Human Tissue Products Liab. Litig., 582 F. Supp. 2d 644, (D.N.J. 2008) (No. 2:06-CV-00135-WJM).

Trotter JF (2003) Transmission of hepatitis C by implantation of a processed bone graft, *J Bone Joint Surg Am* **851**(110), 2215–2217.

Tscherne DM, Jones CT, Evans MJ, Lindenbach BD, McKeating JA, Rice CM (2006) Time and temperature-dependent activation of hepatitis C virus for low pH-triggered entry, *J Virol* **80**(4), 1734–1741.

Tugwell BD, Patel PR, Williams IT, Hedberg K, Chai F, Nainan OV, Thomas AR, Woll JE, Bell BP, Cieslak PR (2005) Transmission of hepatitis C virus to several organ and tissue recipients from an antibody-negative donor, *Ann Intern Med* **143**(9), 648–654.

Uhlenhaut C, Dorner T, Pauli G, Pruss A (2005) Effects of lyophilization on the infectivity of enveloped and non-enveloped viruses in bone tissue, *Biomaterials* **26**(33), 6558–6564.

14

POLYMETHYL METHACRYLATE: AN ACCEPTABLE MATERIAL FOR TISSUE AUGMENTATION AND RECONSTRUCTION

Mario Cabraja, Jack Klecan and Glyn O. Phillips

HISTORY AND MATERIAL DEVELOPMENT

Polymethyl methacrylate has been known scientifically for more than a hundred years. The material was developed in 1928 in various laboratories, and was first brought to market in 1933 by Rohm and Haas Company, under the trademark Plexiglas. The other product names were Perspex (developed by Imperial Chemical Industries (ICI), Runcorn, United Kingdom) and Lucite (developed by DuPont, Philadelphia, United States). However, it was Otto Röhm who first synthesised the polymer polymethyl methacrylate (PMMA) in 1901 and reported the finding in his dissertation at the University of Tübingen (1901). It is a polymer of the monomer methyl methacrylate with the structure shown in Figure 14.1 (Röhm, 1901).

In 1935 ICI introduced an injection moulding technique for dentures in which the molten material was injected under hydraulic pressure into dried plaster moulds. This technique proved to be too cumbersome and critical and was not generally adopted. In 1936 it was discovered by the German Kulzer company that a mixture of methyl methacrylate monomer and ground polymer produced a dough that could be moulded in plaster moulds and could be polymerized to a solid mass by heating using benzoyl peroxide as the polymerization initiator. (Kulzer, 1936).

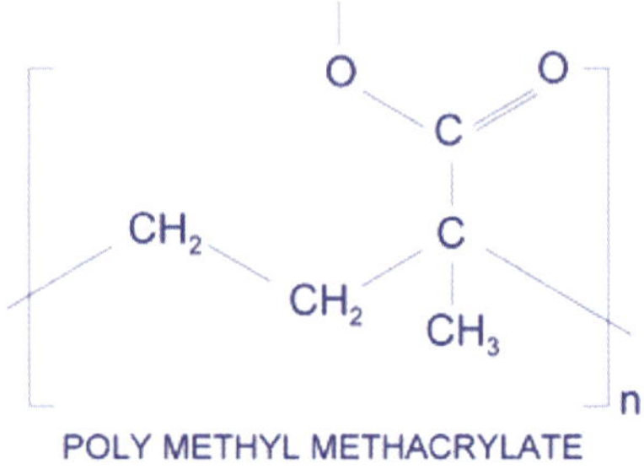

Figure 14.1. Structure of polymethyl methacrylate.

Their development led to the introduction of cold-cured PMMA, which hardens at room temperature. The adoption of the dough moulding technique led to the near universal use of these acrylic resins for dentures and prostheses (e.g., cranioplasty) in the 1940s. It was discovered by German chemists in 1943 that the dough could be polymerized at room temperature if a tertiary amine such as dimethyl-ptoluidine was added with the benzoyl peroxide (Kulzer, 1943). During the Second World War, acrylic glass was used for submarine periscopes, windshields, canopies, and gun turrets for airplanes. This information became generally available at the end of the Second World War in 1945. It was in the dental area that acrylic glass was first applied for medical uses, and considerable research efforts were devoted to producing heat-cured and cold-cured dentures, prosthetics, and filling materials in the 1950s. The properties of acrylic resins for dentistry at that time have been thoroughly reviewed (Tylman and Peyton, 1946; Osborne, 1948; Schwartz, 1950). Acrylic resins have proved to be a consistent and reliable product ever since that time.

Professor Dennis Smith (Smith, 2005) has given a fascinating personal account of how conversations in the Dental School in the University of Manchester led to Professor Sir John Charnley applying PMMA in orthopaedics. Dr Charnley at that time was in the Medical School alongside the University Dental Department and was a frequent visitor there from about 1948. Subsequently, he proved to be the inspirational father of this field. The entire story evokes memories of the eminence and creativity of Manchester University at that time — the alma mater of the great Ernest Rutherford and Robert Robinson.

Charnley was well aware of the range of materials used in dentistry, because his father was a dentist. Although Smith and Charnley were aware of the use of heat-cured acrylics for cranioplasty because such prostheses had been made in the Department, they had not heard about the work of Kiaer (1953), who in 1951 had cemented in six Judet prostheses, nor that of Haboush (1953), who had used acrylic cement as a seating compound on the cut surfaces of the femoral neck. Nor at that time did they know of the use of cold-curing acrylic material in cranioplasty by Spence (1954), Blaine and Oliver (1952), Woringer *et al.* (1951), and Robinson and MacAlister (1954); intercranial aneurysms and spinal fixation by Dutton (1956), Harris (1961), and others. In those days communication was slow and access to literature was limited. Nevertheless these applications were later useful support for demonstrating the safety of acrylic cement in low-friction arthroplasty. Ironically, researchers did not discover until a good many years later that Gluck in Germany had cemented total hip and knee replacements in 1890. Gluck (1891) used ivory components and a cement stated to be composed of colophony resin, pumice, and plaster of Paris. The degree of success, however, is unclear from his article.

Charnley operated on the first case in Manchester in 1958. He reported the preliminary results of six cases (Charnley, 1960). It was a new technique and this provided the basis for the development of Charnley's concept of low-friction arthroplasty (Charnley *et al.*, 1961) over the next decade. Indeed, it was Charnley who is responsible for the initiative which led to the spectacular modern success and popularity of PMMA in orthopaedics.

Information from this dental experience, studies on sensitivity (Smith and Bains, 1956), animal studies (Wiltse *et al.*, 1957), and the later reported experience in neurosurgery (Harris, 1961) showed even at that early stage that the risks from methyl methacrylate were extremely small.

PMMA orthopaedic applications grew exponentially and the excitement of this period can be felt from the monograph published in 1970 setting out Charnley's experiences. Even after five decades, the Charnley system remains the "gold standard" when comparisons are undertaken (Charnley, 1970). Credit must go also to Jean and Robert Judet and Kiaer and Haboush, who separately reported using PMMA to affix femoral implants in the early 1950s (Haboush, 1953). The surgical applications of

PMMA moved quickly from the 1960s, first in hip and disc replacements and then, by the late 1980s, in vertebroplasty and the fixation of spinal fractures.

The first spinal disc replacement was published in 1955 by David Cleveland. He injected PMMA into the disc spaces of 14 patients at the time of discectomy. This procedure yielded "acceptable" results. Cleveland developed his concept of disc replacement in the 1950s after visiting James Gardner, who had replaced hundreds of damaged lumbar discs with Lucite pegs. In 1957, Wallace Hamby presented his results of disc replacement with PMMA at the American Association of Neurological Surgeons (Hamby and Glaser, 1959). He compared disrectomy alone with disrectomy with PMMA and found no difference in terms of hospital stay, return to work, back pain or maintenance of disc height one year after surgery. In 2001, Hamburger *et al.* (1976) published a study of injection of PMMA at the time of anterior cervical disrectomy and fusion with a mean follow-up of 12 years. This procedure yielded 77% "good and excellent" results.

PMMA: PREPARATION AND COMPOSITION

Chemically, PMMA is a synthetic polymer of the monomer methyl methacrylate. Its shape, as shown in Figure 14.2, is stiff-linear, and is responsible for its strength and stiffness.

Dr Todd Jaeblon (Jaeblon, 2010) has contributed a most valuable review of the properties of PMMA, and is the authoritative reference on the PMMA material in relation to its requirements for use in orthopaedics. Material from his work will be referred to in this chapter.

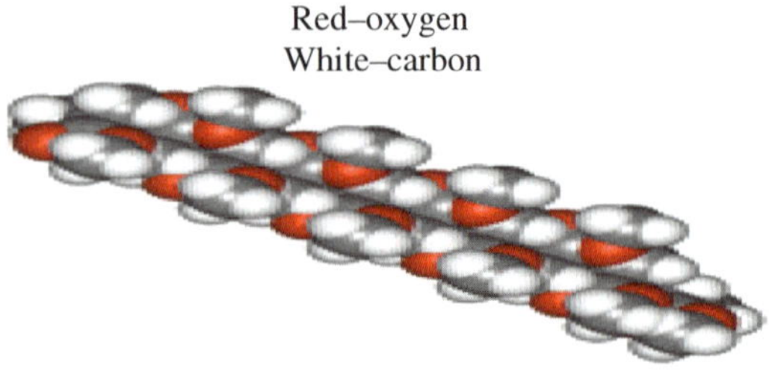

Figure 14.2. Shape of polymethyl methacrylate.

The orthopaedic product is composed of polymer powder and monomer liquid, usually in a 2:1 ratio. The monomer, a colourless liquid with a characteristic odour, is packaged in ampoules. The liquid components remain relatively constant among commercially available cements. Methyl methacrylate makes up 97% to 99% of the liquid; N,N-dimethyl-p-toluidine acts as an accelerator, making up 0.4% to 2.8% by weight. Traces of hydroquinone (15 to 75 ppm) stabilise the monomer to prevent premature polymerisation. Microspheres of ground PMMA or copolymer contribute to 83% to 99% of the powder. The remaining components include a radiopacifier, either barium sulfate ($BaSO_4$) or zirconium dioxide (ZrO_2) (8% to 15% by weight), as well as an initiator, benzoyl peroxide (0.75% to 2.6%). Other variations include the initiator tri-n-butylborane and accelerator 2,5-dimethylhexane-2,5-hydroperoxide (in Bonemite [Mochida Pharmaceutical, Tokyo, Japan]), chlorophyll dye (in the monomer of Palacos R [Zimmer, Warsaw, IN]), and ethanol and ascorbic acid (in the monomer of CMW [DePuy, Piscataway, NJ]) (Lewis, 2008). Other additives to the powder may include antibiotics or dyes.

Combining powder and liquid monomer initiates an exothermic reaction. Peak temperatures in vitro reach 113°C in the anterior cortex of vertebral bodies (Belkoff, 2003). In vivo temperatures are reported to be between 40°C and 56°C (Webb and Spencer, 2007). Methyl methacrylate monomer, the basic building block of PMMA, contains carbon–carbon double bonds, which react with the free radical produced by the activator and initiator. The monomer is free to interact with other monomer molecules, creating a growing polymer chain. The powder initiates polymerisation and creates a workable dough.

Curing is divided into four phases: mixing, sticky, working and hardening. The mixing phase ends once a homogenous state has been achieved. The sticky phase is distinguished by low viscosity, in which the mixture fails to separate from a gloved finger. In the working phase, the cement can be handled without adherence. The hardening phase is the time in which cement cannot be mixed and forms a solid. For most cements, hardening occurs within 10 to 20 minutes (Goodman, 2005; Hass *et al.*, 1975). Commercial PMMA can be categorized as high or low viscosity based on which phase predominates during the curing process. Low-viscosity cements have longer sticky phases and shorter working phases.

High-viscosity cements have long handling times and a short sticky phase. Various formulations are selected according to the location and conditions prevailing.

STANDARDS AND TESTING FOR STABILITY AND STRENGTH

Mechanical, Physical and Stability Standards

A consensus of minimal standards for testing and material performance has been developed for acrylic bone cements by the American Society for Testing and Materials (ASTM) and the International Organization for Standardization (ISO) (Nottrott *et al.*, 2008).

The bending modulus of PMMA is between 1 and 3 GPa, (Harper and Bonfield, 2000; Miller *et al.*, 2007; Kuehn *et al.*, 2005) while that of cortical and cancellous bone is between 10 and 20 GPa and 10 and 2,000 MPa, respectively. PMMA has a compressive strength of between 85 and 110 MPa, compared with 133 to 193 MPa for cortical bone. The tensile strength of commercial PMMA and cortical bone are 30 to 50 MPa (Harper and Bonfield, 2000) and 51 to 133 MPa, respectively (Miller *et al.*, 2007; Kuehn *et al.*, 2005). Fatigue strength, creep, and stress relaxation may be more relevant than tensile strength to long-term clinical performance. Fatigue strength is a product of continuous cyclic loading, creep represents deformation under constant load and stress relaxation represents the changes of stress during constant strain.

A major use of acrylic bone cement is the stable fixation of endoprostheses. The cement must endure considerable stresses. Therefore, sufficient strength is one of the most important demands to achieve a stable fixation and to guarantee long-term stability of the implant. The cement layer has the effect of an elastic buffer between prosthesis and bone (Kuehn *et al.*, 2005).

The compression test according to the standards ISO 5833 and ASTM F 451 is a static method in which the compressive strength is defined as the maximum stress that a material can withstand before failure in compression. The minimum requirement for the compressive strength is 70 MPa according to the standards. All commercial antibiotic and plain bone cements meet this requirement. No significant differences in compressive strength between both cement types are observable. That means the addition of antibiotics does not lower significantly the compressive strength.

The second mechanical test according to ISO 5833 is the four-point bending test, also performed with a universal testing machine. The rectangular specimens have a size of 3.3 × 75.0 × 10.0 mm. The tests are continued until failure and the maximum force is used to calculate the bending strength. Figure 14.3 shows the bending strength and Figure 14.4 shows the bending modulus of several bone cements. According to ISO 5833, the minimum requirement for the bending strength is 50 MPa and for the bending modulus it is 1800 MPa, respectively. Again, all commercial cements clearly fulfil the requirements. The addition of antibiotics reduces the bending strength, but the differences between antibiotic and plain cement are not always statistically significant.

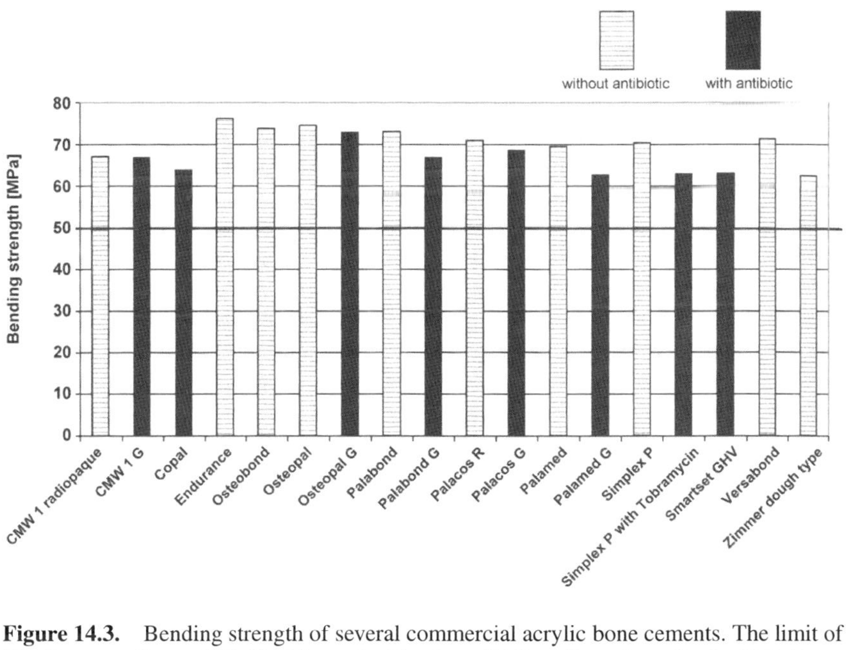

Figure 14.3. Bending strength of several commercial acrylic bone cements. The limit of 50 MPa according to ISO 5833 is marked by the solid line. (Data from Kuehn K-D. Bone cements. Up-to-date comparison of physical and chemical properties of commercial materials, 2000. Heidelberg: Springer Verlag, pp. 141–148, 226–262.)

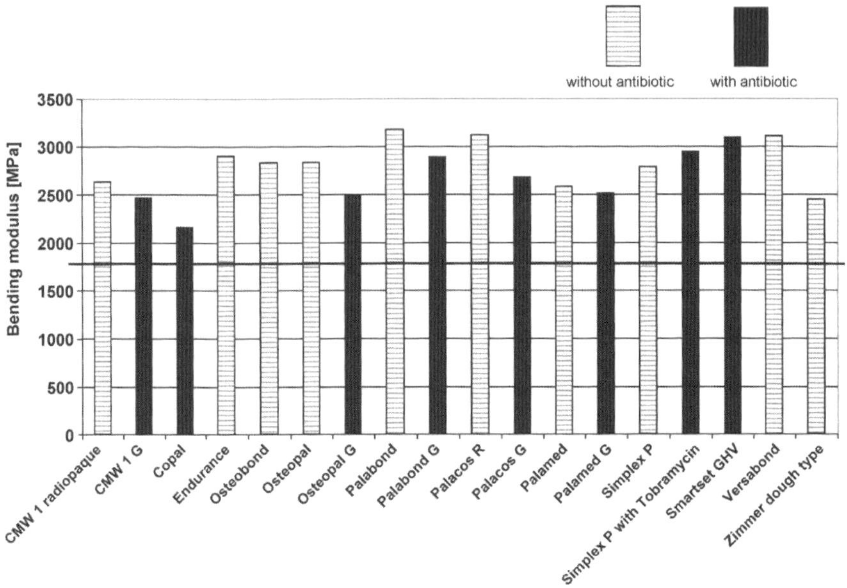

Figure 14.4. Bending modulus of several commercial acrylic bone cements. The limit of 1800 MPa according to ISO 5833 is marked by the solid line. (Data from Kuehn K-D. Bone cements. Up-to-date comparison of physical and chemical properties of commercial materials, 2000. Heidelberg: Springer Verlag, pp. 141–148, 226–262.)

Since the cement acts as a mechanical buffer, the modulus of elasticity of bone cement has to be lower than the moduli of the metallic prosthesis and the bone. The modulus must not decrease to below a minimum value. Therefore, a lower limit for the modulus is established in ISO 5833. The modulus varies with temperature: the higher the temperature, the lower the modulus. Although bone cement has a high compressive strength, it is susceptible to fracture that might result from tensile loading (Harper *et al.*, 2000). Tensile tests therefore are performed according to ISO 527-1 or ASTM D 638.

One further static test applied with PMMA bone cements is the shear strength test according to ASTM D732. This mechanical parameter is important because debonding of the stem — cement interface could have been implicated in the initiation of failure of cemented femoral stems. The interface static shear strength is influenced by surface roughness, cement type and porosity. Surface finish has the greatest effect on the interface strength. Increasing the surface roughness increases the interface shear strength. Increasing the surface roughness to greater than a certain value, however, has no additional effect. Cement type and porosity have a minor influence on the static interface strength only (Wang *et al.*, 2003). Furthermore, there are methods to determine fracture properties, such as fracture toughness (ASTM E399 and ISO 13,586) and impact strength (ISO 179/ISO 180/DIN 53,435) (Sih and Berman, 1980). There is a correlation between the two methods (Lewis and Mladsi, 2000).

The Standards Set by US Food and Drug Administration

The US Orthopedic Devices Branch (ORDB) adopts strict criteria to evaluate PMMA before it can be used clinically. Demian and McDermott (1998) set out the minimum pre-clinical data which must be provided about PMMA, which can include physical and chemical methods of analyses, mechanical test methods, and risks and failure mechanisms. This is part of the responsibility of the US Food and Drug Administration (FDA) as set out in the Medical Device Amendments of 1976 to assure that medical devices are safe, effective and properly labelled for their intended uses. Each medical device is classified into one of three categories:

(1) Class I devices (general controls);
(2) Class II (special controls);
(3) Class III (premarket approval — special controls, labelling requirements, postmarket surveillance and sometimes training requirements).

Even in this relatively early review the authors who were employed by FDA concluded that "the use of acrylic bone cement has enjoyed good clinical success for the past 30 years". PMMA is now categorised as a medical device in Class II. The current requirements are those published on 17 July 2002 by FDA which supersedes the previous 2001 document.

The view of FDA about the safety and effectiveness which gives specific approval for the clinical use of PMMA is given in this conclusion:

> "FDA believes that special controls, when combined with the general controls, will be sufficient to provide reasonable assurance of the safety and effectiveness of PMMA bone cement. Thus, a manufacturer who

> intends to market a device of this generic type should (1) conform to the general controls of the Federal Food, Drug & Cosmetic Act (the Act), including the 510(k) requirements described in 21 CFR 807 Subpart E, (2) address the specific risks to health associated with PMMA bone cement identified in this guidance and, (3) obtain a substantial equivalence determination from FDA prior to marketing the device, unless exempt from the premarket notification requirements of the Act (refer to 21 CFR 807.85)."

The requirements are thus extremely strict and detailed, which ensures that PMMA use is subjected to detailed assessment of its chemical composition, physical and chemical properties, mechanical testing, shelf life, sterility and method of labelling. In parallel, the previously cited ASTM and ISO (Nottrott *et al.*, 2008) must be adhered to.

In view of these extremely rigorous requirements, it is evident that the PMMA used in clinical practice meets the highest standards which are demanded by FDA, ASTM and ISO. Its general approval for use in orthopaedics was extended also to be used in the spine in 2004, specifically for vertebral compression. The conclusion of Kuehn *et al.* (2005) further supports the excellence of this well-characterised and tested surgical device: "The cemented hip arthroplasty is one of the most successful orthopaedic operations performed today, and acrylic bone cement is the most commonly used non-metallic implant material in orthopaedics."

GENERAL CLINICAL UTILISATION

The function of PMMA in arthroplasty is to fill voids left by mismatches between host bone and implant, thus creating immediate stability. The transfer and distribution of forces from implant to bone is thereby subject to a more physiologic transition as a result of the mechanical properties of PMMA, which approximate those of bone. PMMA also dampens excessive forces that would otherwise be directly applied to host bone. PMMA has no adhesive properties to implants at a molecular level. It is dependent on surface properties and shape to enhance stability (Crowninshield, 2001). The quality of apposition between the implant-cement and bone-cement interfaces is of paramount importance in determining the longevity of a cemented prosthesis. These interfaces are directly or indirectly affected by surgical technique and loading characteristics as well as by the properties of cement, bone and implant. Gravius *et al.* (2008) reported fewer cement mantle cracks and gap defects with the use of femoral stems that are anatomically formed, collared and well rounded. An anatomic stem results in a uniform cement mantle, while the collar decreases tensile stresses on the mantle.

PMMA has been used successfully in orthopaedic surgery for more than 60 years. Over this time its compatibility and lack of toxicity with human tissue have been demonstrated in a wide series of investigations over the period 1973–1995. (Whitehill *et al.*, 1988; Branch *et al.*, 1989; Rubin, 1983; Cheung *et al.*, 1994; Schwartz *et al.*, 1973; Govila, 1990; Benoist, 1978; van Mullem and de Wijn, 1988; Wellisz *et al.*, 1995; Worley, 1973; Al-Sibahi and Shanoon, 1982).

The use of PMMA is now the preferred procedure worldwide for use with prostheses implantation. Hailey and Turner (1994) state, "Total joint replacement is a procedure which gives pain relief and renewed mobility to over 50,000 people each year in the UK alone." Within the United States, over 600,000 total hip and total knee replacements are performed each year (The Kaiser Permanente National Total Joint Replacement Registry) and "more than 800,000 hip replacement operations are carried out worldwide every year" (McCormack *et al.*, 1998). By the year 2030, that number is projected to exceed 4 million (Kurtz *et al.*, 2007).

There have also been major international evaluations of the use of PMMA in total hip replacement. The surgical practice is widely affirmed and supported by registry recorded data (Furnes *et al.*, 2005; Hook *et al.*, 2006; Malchau *et al.*, 2002; Williams *et al.*, 2002; Fowler *et al.*, 1988). The bio-mechanical characteristics of PMMA are key to this success.

Literally thousands of orthopaedic applications of PMMA have been reviewed without serious adverse results. Ahnfelt *et al.* (1990) reviewed 4,664 total hip replacements using PMMA with good results overall. These results just add to the mass of clinical data showing the widespread use of PMMA and the success of the clinical outcome.

Goodger *et al.* (2005) demonstrated that PMMA is well tolerated in oral surgery. They demonstrated that PMMA blocks allow the overlying oral mucosa and skin to heal without delay or wound breakdown. Their conclusion is that "PMMA is reported to be well tolerated by bone and soft tissues, and this has been confirmed in our experience".

The investigations reported in the following papers support the wide-ranging research that PMMA is biocompatible and well tolerated by tissues of the human body. This is the best possible testimony derived from actual surgical interventions and then evaluated over many years for possible problems. The results are a resounding endorsement for the international use of PMMA in orthopaedic surgery and its complete tolerance by the human body.

Garellick *et al.* (1994) carried out a retrospective, long-term study to analyse clinical function, failures and radiographic status in 95 Charnley low-friction arthroplasties performed during the period 1973–1977. They found that "ninety four percent of the patients were satisfied with their results". The clinical conclusion was, "We think that our earliest LFAs have performed remarkably well after 12–16 years. Cemented hip arthroplasty continues to be the treatment of choice in older and relatively inactive patients."

Gautschia *et al.* (2010) more recently have concluded that "PMMA is characterised by excellent biocompatibility with low intrinsic toxicity, inflammatory activation and immunogenicity, resulting in a good tolerance by the majority of patients". They report on the only one ever-reported case in the literature of hypersensitivity to bone cement cranioplasty.

These clinical investigations of the safely and effectiveness of PMMA have been detailed and extremely thorough, but for specific details the original publications should be directly consulted. A specific study by Jensen *et al.* (1997) reviewed the application of PMMA for the treatment of osteoporotic vertebral body compression fractures. The conclusion based on 29 patients with 47 painful vertebral fractures is the following: "Complications appear to be rare. Vertebroplasty (with PMMA) is a valuable new tool in the treatment of osteoporotic vertebral compression fractures, providing pain relief and early mobility in appropriate patients."

Another study to evaluate the safety and efficacy of cement augmentation was undertaken by Lieberman *et al.* (2001) for the treatment of painful osteoporotic vertebral compression fractures. Seventy consecutive kyphoplasty procedures were performed in 30 patients. There were no major complications observed. In 70% of the vertebral bodies the treatment restored 47% of the lost height, and it was concluded that the procedure "was efficacious in the treatment of osteoporosis vertebral compression". All 30 patients tolerated the procedure well and improvement in pain and mobility was seen early. Virtually all patients reported immediate relief of their typical fracture pain and no patient complained of worse pain. The authors state that the efficacy study suggests that "kyphoplasty is a safe procedure".

A more recent publication (Cabraja *et al.*, 2011) has again supported the long-term safety and efficacy of PMMA. This study presents a long-term comparison of clinical and radiological data from patients with pseudarthrosis and solid arthrodesis after PMMA-assisted ventral discectomy was performed. From 1986 to 2004. 416 patients underwent ventral discectomy and PMMA interposition. The clinical and radiological outcome was assessed for 50 of 127 eligible patients after a mean of 8.1 years. Based on postoperative radiographs, the patients were dichotomised into those with pseudarthrosis (group A) and those with solid arthrodesis (group B). Pseudarthrosis with movement of more than two of the operated segments was noted in 17 cases (group A). In 33 cases no movement of the vertebral segment could be detected (group B). The overall conclusion is that PMMA-assisted discectomy shows a high rate of pseudarthrosis. But the clinical long-term success does not seem to be negatively affected by this.

Due to the comparably low fusion rates after intervertebral placement of PMMA in the cervical spine, many authors recommended the use of PMMA only in cases without the explicit need for bony fusion. This condition is especially required following a trauma. However, in the field of surgical treatment of degenerative disc disease, bony fusion is not mandatory.

In the German-speaking countries PMMA was widely accepted as a safe, reliable and economic substrate for intervertebral disc replacement in the cervical spine with decades of positive experience. Several techniques had been developed to avoid a graft dislocation and injury of nerval structures (see Figure 14.5). Even today PMMA is used in a number of German neurosurgical departments to replace the removed disc (Topalovic *et al.*, 2013).

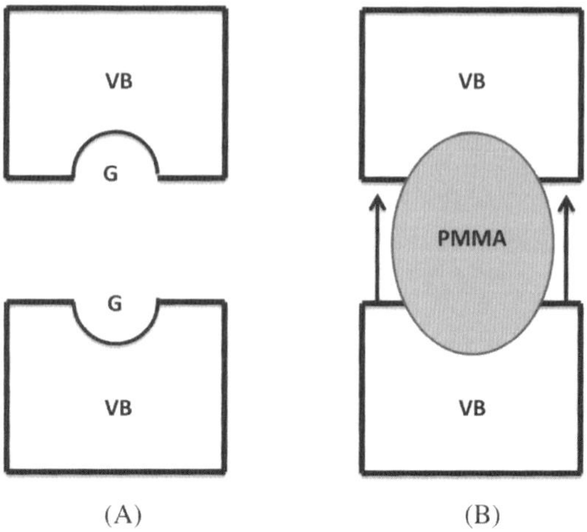

Figure 14.5. Insertion of PMMA in the intervertebral disc space of the cervical spine: (A) After disc removal grooves (G) were created into the vertebral bodies (VB). (B) PMMA was inserted into the grooves to avoid dislocation. The intervertebral space was not filled entirely to enable a bone growth and subsequent fusion (arrows). The space facing the dura mater (arrow) was also used as a "buffer" to protect nerval structures from thermal injury by previous placing of gel foam between the hardening PMMA graft and the dura mater.

Numerous studies supported the positive experience with application of PMMA in the cervical spine. Topalovic *et al.* (2013) reviewed 12 studies in the past 30 years that investigated PMMA-assisted surgery of the cervical spine. The reviewed studies covered up to 20 years and proved the clinical long-term success of PMMA in more than 1,600 treated patients. Five of these studies had a prospective study design. Two studies prospectively compared PMMA insertion into the intervertebral disc space after disc removal with titanium cages and autologous bone graft, both substrates being the most widely accepted materials to replace the removed disc (Bärlocher *et al.*, 2002; Schröder *et al.*, 2007). As expected, PMMA could not offer equal ability to fuse the intervertebral space, but offered equally good clinical outcomes and proved to be a safe and reliable disc substitute in the treatment of the degenerative disc disease. It also offered certain advantages compared to autologous bone graft and titanium cages: application of PMMA avoids the problem of donor site morbidity compared with the former and reduces costs compared to the latter without compromising clinical long-term success. The cost factor of synthetic cages triggered the development of cheaper PMMA-cages that could be filled with bone and thus improve the graft management, shorten time of surgery and reduce the rate of non-fusion substantially (Chen *et al.*, 2005). A prospective study (Chen *et al.*, 2009) was performed with 92 consecutive patients who underwent single-level anterior cervical discectomy and fusion (ACDF) with a hollow PMMA spacer, cancellous allograft and titanium cervical plate stabilisation between January 2002 and December 2003. Patients were followed for a minimum of two years. The conclusion: "The procedure is safe and effective. There were no complications related to the hollow PMMA spacer. This procedure has a high fusion rate, and can restore disc height and maintain normal cervical lordosis. This method achieves results similar to those of other methods."

For any implanted material into the human body it is important that there is biocompatibility with blood and components. There is evidence that the extract of well-formed methacrylate-based bone cement, i.e., saline or buffer incubated with bone cement for 72 hours at 37°C, did not affect coagulation tests, had no haemolytic activity and did not influence tissue factor production or thrombomodulin levels in cultured endothelial cells (Cenni *et al.*, 2001a; 2001b; 2001c).

Blinc *et al.* (2004) carried out a further detailed investigation into the effects on blood and blood components. This study again has further dem onstrated the compatibility which PMMA has with the material which it comes into contact with after surgical implantation.

These researchers demonstrated that the surface of fresh or aged bone cement is not thrombogenic in vitro. They state in conclusion that "our study adds to the body of evidence that bone cement used in vertebroplasty is rather safe". Of course, the key to avoiding complications during vertebroplasty is proper technique of polymethyl-methacrylate delivery into the vertebral body with fluoroscopic visualisation and low-pressure injection of adequate volume as previously shown by Moreland *et al.* (2000).

At an early stage (1988) it was confirmed that there is no cancerous effect due to the PMMA polymer. This has been borne out by experience subsequently. However, it is important to have on record the findings of Chan *et al.* (1988) who demonstrated that there was no evidence of carcinogenicity due to PMMA for male rats exposed by inhalation at 500 or 1000 ppm, for female rats at 250 or 500 ppm, or for male and female B6C3F1 mice at 500 or 1000 ppm.

It should be noted that PMMA is an internationally recognised quality product with established chemical, physical, mechanical and clinical properties. Literally thousands of physical and medical applications of PMMA are annually reported in respected scientific journals.

The uses of PMMA now apply to almost all parts of the body where stability and augmentation of the damaged or diseased structures are required. A review of its performance over decades confirms its long-term effectiveness. As illustration, a review of lens implants after cataract surgery over 20 years ago, skull implants, cranioplasty and vertebral stabilization after insertion during the period 2003–2010, use for vertebro-plasty, the percutaneous augmentation of vertebral bodies since the late 1980s, have all proved positive. (Case Report: 2012; Bos and Tinga, 2013; Socola *et al.*, 2010; Thielena *et al.*, 2009; Galibert *et al.*, 1987).

The physical material has mainly been used as prepared *in situ*. However, it has also been used in association with other biocompatible materials such as extruded high-quality film or sheet, osteoinductive hydroxyapetite coatings, metal implants and/or antibiotics. For example, intraoperatively manufactured, antibiotic-releasing PMMA hip interim prostheses with titanium endoskeleton allow for flexibility in choice of antibiotic and geometry. A suitable spacer is capable of fulfilling patients' needs.

Today PMMA is the most commonly used surgical augmentation material. Its material qualities are of such a high quality that now there are also a multitude of applications outside the medical field such as for injection moulded baths, sinks and kitchen worktops, automotive cluster lenses, car rear lights, composites, dials, Fresnel lenses, household items, light guides, optical fibres, point-of-purchase displays, rods and solar collector lenses (Thielena *et al.*, 2009; Socola *et al.*, 2010; Chem Tec Publishing, 2012).

REFERENCES

Ahnfelt L, Herberts P, Malchau H, Andersson GB (1990) Prognosis of total hip replacement: A Swedish multicenter study of 4,664 revisions, *Acta Orthop Scand. Suppl.* **238**, 1–26.

Al-Sibahi A, Shanoon A (1982) The use of soft polymethyl methacrylate in the closure of oro-antral fistula, *J Oral Maxillofac Surg* **40**, 165.

Bärlocher CB, Barth A, Krauss JK, Binggeli R, Seiler RW (2002) Comparative evaluation of microdiscectomy only, autograft fusion, polymethyl methacrylate interposition, and threaded titanium cage fusion for treatment of single-level cervical disc disease: A prospective randomized study in 125 patients, *Neurosurg Focus* **12**(1), E4.

Belkoff SM, Molloy S (2003) Temperature measurement during polymerization of polymethyl methacrylate cement used for vertebroplasty, *Spine (Phila Pa 1976)* **28**, 1555–1559.

Benoist M (1978) Experience with 220 cases of mandibular reconstruction, *J Maxillofac Surg* **6**, 40.

Blaine G, Oliver LC (1952) Acrylic cranioplasty, *Br J Surg* **39**, 371–372.

Blinc A, Bozic M, Vengust R, Stegnar M (2004) Methyl-methacrylate bone cement surface does not promote platelet aggregation or plasma coagulation in vitro, *Thrombosis Research* **114**, 179–184.

Bos R, Tinga R (2013) Success rate of custom made PMMA implants for cranial reconstruction: A follow up study and comparison with the regulatory perspective on characterization and testing literature 21st ICOMS 2013 — Abstracts: Oral Papers, University Medical Center Groningen, The Netherlands.

Branch CL, Kelly DL, Davis CH Jr, *et al.* (1989) Fixation of fractures of the lower cervical spine using methyl methacrylate and wire: Technique and results in 99 patients, *Neurosurgery* **25**, 503.

Cabanela ME, Coventry MB, MacCarty CS *et al.* (1972) The fate of patients with methyl methacrylate cranioplasty, *J Bone Joint Surg* **54**, 278.

Cabraja KD, Lanksch WR, Maier-Hauff K, Kroppenstedt S (2011), Polymethyl methacrylate-assisted ventral discectomy: Rate of pseudarthrosis and clinical outcome with a minimum follow-up of 5 years, *BMC Musculoskeletal Disorders* **12**, 140.

Cenni E, Ciapetti G, Granchi D, Savarino L, Stea S, Corradini A *et al.* (2001a) Evaluation of tissue factor production by human endothelial cells incubated with three acrylic bone cements, *J Biomed Mater Res* **55**, 131–136.

Cenni E, Ciapetti G, Granchi D, Savarino L, Stea S, Corradini A (2001b) Evaluation of the effect of seven acrylic bone cements on erythrocytes and plasmatic phase of coagulation, *Biomaterials* **22**, 1321–1326.

Cenni E, Ciapetti G, Granchi D, Stea S, Savarino L, Corradini A *et al.* (2001c) No effect of methacrylate-based bone cement CMW 1 on the plasmatic phase of coagulation, red blood cells and endothelial cells in vitro, *Acta Orthop Scand* **72**, 86–93.

Charnley J (1960) Anchorage of the femoral head prosthesis to the shaft of the femur, *J Bone Joint Surg* **42B**, 28–30.

Charnley J (1961) Arthroplasty of the hip by the low friction technique, *J Bone Joint Surg* **43B**, 601.

Charnley J (1970) *Acrylic Cement in Orthopaedic Surgery*, E & S Livingstone, Edinburgh.

ChemTec Publishing (2012) *Applications of PMMA, Handbook of Polymers*, p. 454.

Chen JF, Wu CT, Lee ST (2009) Spacer with plating in the treatment of single level cervical disc disease, *Chang Gung Med J* **32**, 447–454.

Chen JF, Wu CT, Lee SC, Lee S (2005) Use of a polymethyl methacrylate cervical cage in the treatment of single-level cervical disc disease, *J Neurosurg Spine* **3**(1), 24–28.

Cheung LK, Samman N, Tideman H (1994) The use of moldable acrylic for restoration of the temporalis flap donor site, *J Craniomaxillofac Surg* **22**, 335.

Cleveland DA (1955) Management of cerrical disk and cervical arthritis syndromes, *Postgrad Med* **18**, 99–105.

Crowninshield R (2001) Femoral hip implant fixation within bone cement, *Oper Tech Orthop* **11**, 269–299.

Demian HW, McDermott K (1998) Regulatory perspective on characterization and testing of orthopedic bone cements (US Food and Drug Administration Center for Devices and Radiological Health, Office of Device Evaluation), *Biomaterials* 19, 1607–1618.

Dutton J (1956) Intercranial aneurysm: A new method of surgical treatment, *Br Med J* **8**, 585.

FDA Class II Special Controls Guidance Document: Polymethyl methacrylate (PMMA) Bone Cement; Guidance for Industry and FDA Document issued on July 17, 2002; U.S. Department of Health and Human Services; Food and Drug Administration; Center for Devices and Radiological Health; Orthopedic Devices Branch; Division of General, Restorative, and Neurological Devices; Office of Device Evaluation.

Fowler JL, Gie GA, Lee AJ, Ling RS (1988) Experience with the Exeter total hip replacement since 1970, *Orthop Clin North Am* **19**, 477–489.

Furnes O, Espehaug B, Lie SA *et al*. (2005) Prospective studies of hip and knee prostheses. The Norwegian Arthroplasty Register 1987–2004. American Academy of Orthopaedic Surgeons, 23–2.

Galibert P, Deramond H, Rosat P, Gars DL (1987) Preliminary note on the treatment of vertebral angioma by percutaneous acrylic vertebroplasty, *Neurochirurgie* **33**(2), 166–168.

Garellick G, Herberts P, Striimberg C, Malchau H (1994) Long-term results of Charnley arthroplasty: A 12–16-year follow-up study, *The Journal of Arthroplasty* **9**(4), 333–340.

Gautschia OP, Schlettb CL, Fournier JV, Cadosch D (2010) Case report: Laboratory confirmed polymethyl-methacrylate (Palacos®)-hypersensitivity, after cranioplasty, *Clin Neurol Neurosurg* **112**, 915–916.

Gluck T (1891) Referat über die durch das modern chirurgische Experiment gewonnenen positiven Resultate, betreffend die Naht und den Ersatz von Defecten höherer Gewebe, sowie über die Verwertung resorbierbarer und lebendiger Tampons in der Chirurgie, *Arch klin chir* **41**, 187–239.

Goodger NM, Wang J, Smagalski GW, Hepworth B (2005) Methyl methacrylate as a space maintainer in mandibular reconstruction, *J Oral Maxillofac Surg* **63**, 1048–1051.

Goodman S (2005) Wear particulate and osteolysis, *Orthop Clin North Am* **36**, 41–48, vi.

Govila A (1990) Use of methyl methacrylate in bone reconstruction, *Br J Plast Surg* **43**, 210.

Gravius S, Wirtz DC, Siebert CH *et al.* (2008) *In vitro* interface and cement mantle analysis of different femur stem designs, *J Biomech* **41**, 2021–2028.

Haas SS, Brauer GM, Dickson G (1975) A characterization of polymethyl methacrylate bone cement, *J Bone Joint Surg Am* **57**, 380–391.

Haboush EJ (1953) A new operation for arthroplasty of the hip based on biomechanics, photoelasticity, fastsetting dental acrylic, and other considerations. *Bull Hosp Joint Dis* **14**, 242–277.

Hailey JL, Turner IG, Miles AW, Price G (1994) The effect of post-curing chemical changes on the mechanical properties of acrylic bone cement, *J Mater Sci Mater Med* **5(9–10)**, 617–621.

Harper EJ, Bonfield W (2000) Tensile characteristics of ten commercial acrylic bone cements, *J Biomed Mater Res* **53**, 605–616.

Harris P (1961) Spinal fixation using onlay of Simplex-P. Paper No. 36. Washington 7 Excerpta Med International Congress Series, p. E81–2.

Hook S, Moulder E, Yates PJ *et al.* (2006) The Exeter universal stem: A minimum ten year review from an independent centre, *J Bone Joint Surg [Br]* **88-B**, 1584–1590.

Jaeblon T (2010) Polymethyl methacrylate: Properties and contemporary uses in orthopaedics, *J Am Acad Orthop Surg* **18**(5), 297–305.

Jensen ME, Evans AJ, Mathis JM, Kallmes DF, Cloft HJ, Dion JE (1997) *Am J Neuroradiol* **18**, 1897–1904.

Kuehn KD, Ege W, Gopp U (2005) Acrylic bone cements: Mechanical and physical properties, *Orthop Clin North Am* **36**, 29–39.

Kulzer Co. DRP Patent 737 058 1936.

Kulzer Co. DRP Patent 973 590 1943.

Kurtz S, Ong K, Lau E, Mowat F, Halpern M (2007) Projections of primary and revision hip and knee arthroplasty in the United States from 2005 to 2030, *J Bone Joint Surg Am* **89**(4), 780–785.

Lewis G (2008) Alternative acrylic bone cement formulations for cemented arthroplasties: Present status, key issues, and future prospects, *J Biomed Mater Res B Appl Biomater* **84**, 301–319.

Lewis G, Mladsi S (2000) Correlation between impact strength and fracture toughness of PMMA-based bone cements, *Biomaterials* **21**(8), 775–781.

Lieberman IH, Dudeney S, Reinhardt MK, Bell G (2001) Initial outcome and efficiency of "kyphoplasty" in the treatment of painful osteoporotic vertebral compression fractures, *Spine* **26**, 1631–1638.

Malchau H, Herberts P, Eisler T, Garellick G, Soderman P (2002) The Swedish total hip replacement register, *J Bone Joint Surg [Am]* **84-A**(Suppl 2), 2–20.

McCormack *et al.* (1998) Mechanical considerations of polymethyl methacrylate in total hip replacement, *Int J Fatigue* **20**(3), 581.

Miller JD, McCreadie BR, Alford AI, Hankenson KD, Golstein SA (2007) Form and function of bone, in Einhorn TA, O'Keefe RJ, Buckwalter JA (eds.) *Orthopaedic Basic Science: Foundations of Clinical Practice*, ed. 3, Rosemont, IL, American Academy of Orthopaedic Surgeons, pp. 129–159.

Moreland DB, Landi MK, Grand W (2000) Vertebroplasty: Techniques to avoid complications, *Spine J* **1**, 66–71.

[No authors listed] (1953) Hip arthroplasty with acrylic prosthesis. *Acta Orth Op Scand* 22(4), 344–350.

Nottrott M, Mølster AO, Moldstad IO, Walsh WR, Gjerdet NR (2008) Performance of bone cements: Are current preclinincal specifications adequate? *Acta Orthop* **79**, 826–831.

Osborne J (1948) *Acrylic Resins in Dentistry*, Blackwell, Oxford.

Robinson RG, MacAlister AD (1954) Acrylic cranioplasty: A simple one-stage method using a cold curing material, *Br J Surg* **42**, 312–315.

Röhm O (1901) On the polymerization, products of acrylic acid [dissertation]. University of Tübingen, Tübingen, Germany.

Rubin LR (1983) Bone cement implantation, in Rubin LR (ed.) Biomaterials in Reconstructive Surgery, CV Mosby, St Louis, MO, p. 452.

Schröder J, Grosse-Dresselhaus F, Schul C, Wassmann H (2007) PMMA versus titanium cage after anterior cervical discectomy: A prospective randomized trial, *Zentralbl Neurochir* **68**(1), 2–7.

Schwartz AB, Kaufman P, Gluck M (1973) Intraoral implantation of methyl methacrylate for the restoration of facial contour in old fractures of the malar complex, *J Oral Surg* **31**, 763.

Schwartz JR (1950) *The Acrylic Plastics in Dentistry*, Dental Items of Interest Publishing Co., Brooklyn.

Sih GC, Berman AT (1980) Fracture toughness concept applied to methyl methacrylate, *J Biomed Mater Res* **14**(3), 311–324.

Smith DC (2005) The genesis and evolution of acrylic bone cement, *Orthop Clin North Am* **36**, 1–10.

Smith DC, Bains MED (1956) The detection and estimation of residual monomer in polymethyl methacrylate, *J Dent Res* **35**, 16–24.

Spence WT (1954) Form-fitting plastic cranioplasty, *J Neurosurg* **11**, 219–225.

Socola G, Macoveib AM, Miroiua F, Stefana N, Dutaa L, Dorciomana G, Mihailescua IN, Petrescu SM, Stan GE, Marcovc DA, Chiriac A, Poeat I (2010) Hydroxyapatite thin films synthesized by pulsed laser deposition and magnetron sputtering on PMMA substrates for medical applications, *Materials Science and Engineering B* **169**, 159–168.

Thielena T, Maas S, Zuerbes A, Waldmann D, Anagnostakos K, Kelm J (2009) Development of a reinforced PMMA-based hip spacer adapted to patients' needs, *Med Eng Phys* **31**(8), 930–936.

Topalovic M, Kroppenstedt S, Cabraja M (2013) The use of polymethyl methacrylate in cervical spine surgery, *Hard Tissue* **2**(3), 24.

Tylman SD, Peyton FA (1946) *Acrylics and Other Synthetic Resins Used in Dentistry*, JB Lippincott, Philadelphia.

van Mullem PJ, de Wijn JR (1988) Bone and soft connective tissue response to porous acrylic implants, *J Craniomaxillofac Surg* **16**, 99.

Wang JS, Taylor M, Flivik G, Lidgren L (2003) Factors affecting the static shear strength of the prosthetic stem–bone cement interface, *J Mater Sci Mater Med* **14**, 55–61.

Webb JC, Spencer RF (2007) The role of polymethyl methacrylate bone cement in modern orthopaedic surgery, *J Bone Joint Surg Br* **89**, 851–857.

Wellisz T, Lawrence M, Jazayeri MA *et al.* (1995) The effects of alloplastic implant onlays on bone in the rabbit mandible, *Plast Reconstr Surg* **96**, 957.

Whitehill R, Cicoria AD, Hooper WE *et al.* (1988) Posterior cervical reconstruction with methyl methacrylate cement and wire: A clinical review, *J Neurosurg* **68**, 576.

Williams HD, Browne G, Gie GA *et al.* (2002) The Exeter universal cemented femoral component at 8 to 12 years: A study of the first 325 hips, *J Bone Joint Surg [Br]* **84-B**, 324–334.

Wiltse LL, Hall RH, Stenehjem JC (1957) Experimental studies regarding the possible use of self-curing acrylic in orthopaedic surgery, *J Bone Joint Surg Am* **39A**(4), 961–972.

Woringer E, Schweig B, Brogly G, Schneider J (1951) Nouvelle technique ultra-rapide pour la refection de breches osseuses craniennes a la resine acrylique: Advantages de la resine acrylique sur la tantale, *Rev Neurol* **85**, 527–535.

Worley RD (1973) The experimental use of polymethyl methacrylate implants in mandibular defects, *J Oral Surg* **31**, 170.

INDEX